Oxford Chemistry Series

General Editors
P. W. ATKINS   J. S. E. HOLKER   A. K. HOLLIDAY

Oxford Chemistry Series

COLIN F. BELL
SENIOR LECTURER IN CHEMISTRY, BRUNEL UNIVERSITY

Principles and applications of metal chelation

Clarendon Press · Oxford · 1977

Oxford University Press, Walton Street, Oxford OX2 6DP

GLASGOW NEW YORK TORONTO MELBOURNE WELLINGTON
CAPE TOWN IBADAN NAIROBI DAR ES SALAAM LUSAKA ADDIS ABABA
DELHI BOMBAY CALCUTTA MADRAS KARACHI LAHORE DACCA
KUALA LUMPUR SINGAPORE HONG KONG TOKYO

ISBN 0 19 855485 0

© Oxford University Press 1977

Printed in Great Britain
by Billing & Sons Ltd., Guildford and Worcester

Editor's foreword

Metal chelates are metal complexes — but with properties which may sharply distinguish them from non-chelate complexes. They are almost invariably organometallic compounds — but again their properties are very different from most of the compounds conventionally regarded as 'organometallic'. They may often be formed by quite simple reactions in aqueous solution; so the relevant equilibria can readily be studied and thermodynamic data obtained. The range and variety of chelating ligands are enormous; so, therefore, are the structures of the resulting complexes, and their importance in biological systems and industrial applications is a further consequence of their rich variety.

This book, then, is concerned with all these characteristics of metal chelates. They form a distinct area topic in chemistry but one which, because it is contingent upon so many other areas of study, should be of interest to all.

A.K.H.

Preface

This book is an introductory account of the interactions between metal ions and chelating ligands, together with a description of some of their more important applications. It is intended for use as a text by students in the later years of a degree course in chemistry or biochemistry. Chelation spans inorganic, organic, and physical chemistry and is of great significance in many branches of analytical chemistry. It is fundamental to the operation of many vital biological processes and has numerous and increasing applications in the field of medicine.

Chelation has such wide ramifications throughout the sciences of chemistry and biology that any introductory treatment must necessarily be highly selective. This book is written from the viewpoint of an inorganic chemist, interested particularly in the properties of metal ions and chelating ligands and the structural aspects of metal chelates and their applications in analysis, biology, and industry. The thermodynamics of chelation reactions are also considered and throughout the book reference is made to physico-chemical data to help the reader towards a more quantitative understanding of metal-ligand interactions.

My interests in chelation were stimulated when I had the good fortune to carry out research work with Professor H. M. Irving on the chemistry of dithizone and some of its metal complexes. His research contributions to our current knowledge of chelate chemistry are too well known to need elaboration here. I would, however, like to take this opportunity to acknowledge gratefully my personal debt to him for his continued help and encouragement in my work.

I wish to express my gratitude to Professor A. K. Holliday for his valuable advice during the writing of this book and to Dr. D. A. L. Hope for his helpful comments on thermodynamic aspects of chelation. I would like also to thank my daughter, Linda, for her great help in the preparation of diagrams and in indexing and finally to express my appreciation to my wife, Sheila, who has once again undertaken the typing of the manuscript so efficiently.

C.F.B.

Contents

Introduction

The nature of chelation

A metal *complex* is formed by association between a metal atom or ion and another species, known as the *ligand*, which is either an anion or a polar molecule. Many complexes are relatively unreactive entities. They remain unchanged throughout a sequence of chemical or physical operations and can often be isolated as stable solid or liquid compounds. Other complexes have a much more transient existence. They may exist only in solution or they may be highly reactive and easily converted to other chemical species. All metals form complexes, although the extent of formation and nature of these depend very largely on the electronic structure of the metal. In the case of transition metals, the phenomenon of complex formation is so widespread that it provides the key to an understanding of the bulk of their chemistry.

The concept of a metal complex originated in the work of Alfred Werner. On the basis of his extensive research into coordination compounds of certain transition metals, he concluded that a metal ion was characterized by two valencies, which he called *principal* and *auxiliary*. The principal valency is now termed the *oxidation state* or *oxidation number* of the metal. The auxiliary valency represents the number of ligand atoms associated with the central metal atom and is therefore the same as the *coordination number* of the metal. For many metals in their lower oxidation states ($+1$, $+2$, $+3$), the coordination number is usually 4 or 6: other values, less commonly observed, are 2, 3, 5, 7, 8, 9, and 10.

Werner synthesized a series of ammonia complexes of cobalt(III). These contained complex ions such as $[Co(NH_3)_6]^{3+}$, $[Co(NH_3)_5Cl]^{2+}$, and $[Co(NH_3)_4Cl_2]^+$: in all three examples, cobalt shows an oxidation number of $+3$ and its coordination number is 6.

A simple description of the nature of the bond between a ligand and a metal treats the ligand as an electron-pair donor and the metal as an electron-pair acceptor. In other words, the metal behaves as a Lewis acid and the ligand as a Lewis base. The donation of a pair of electrons from ligand to metal establishes a coordinate bond. For example, in $[Co(NH_3)_6]^{3+}$, the nitrogen atom of each ammonia molecule carries a lone pair of electrons which it donates to a vacant orbital in the cobalt(III) ion. In the terminology of molecular orbital (MO) theory, orbitals on the ligand and metal of σ-symmetry overlap to give a σ-bonding MO which contains the lone pair of electrons originating on the ligand.

Other factors contribute to metal-ligand bonding in coordination compounds. Thus a major part of the bonding in many complexes comes from the electrostatic attraction between the metal ion and the anionic or polar ligand. In some cases, where electrons and orbitals of appropriate symmetry are available in the metal and ligand, π-bonding makes an additional contribution. Where metal d-electrons interact with vacant ligand orbitals to form a π-bond, this represents a transfer of electronic charge from metal to ligand, that is in the reverse direction to that resulting from σ-bonding.

In the case of ligands such as chloride and ammonia, only one pair of electrons is involved in σ-bonding with a metal. This type of ligand is called *unidentate* (literally 'one-toothed'). Many molecules or ions contain more than one donor atom and it may be sterically possible for them to coordinate one metal atom at two or more positions in its cordination shell — to behave in a *multidentate* fashion. For example, ethylenediamine(1,2-diaminoethane), $NH_2CH_2CH_2NH_2$, contains two donor nitrogens and acts as a *bidentate* ligand towards many metals.

This behaviour of ethylenediamine was discovered in 1893 by Werner in the complex bis(ethylenediamine)platinum(II) chloride, [Pt en$_2$]Cl$_2$. Here the nitrogen atoms of each ethylenediamine molecule occupy two positions in the coordination shell of the platinum atom and the environment of this is exactly the same as in the tetrammine complex, $Pt(NH_3)_4Cl_2$. In both, the coordination number of platinum is four.

In similar fashion, ethylenediamine forms complexes with cobalt(III) ions, for example, dichlorobis(ethylenediamine)cobalt(III) chloride [Co en$_2$Cl$_2$]Cl, one of the first prepared by Werner. He discovered that two isomeric forms of this compound can exist and this, together with other instances of isomerism in cobalt(III) complexes, led to his proposals that the metal has a coordination number of 6 and the donor atoms are situated at the vertices of an octahedron with the metal atom at its centre. Isomerism in [Co en$_2$Cl$_2$]Cl arises because there are two stereochemically different arrangements of the coordinated chlorines, either in a *cis-* or a *trans*-disposition to one another.

tetrammineplatinum(II)

bis(ethylenediamine)platinum(II)

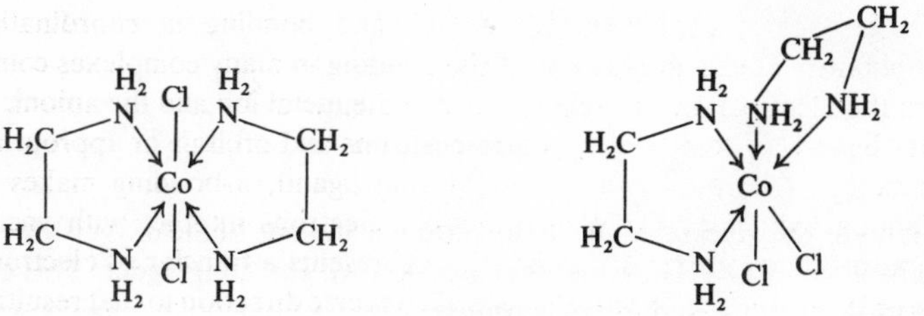

dichlorobis(ethylenediamine)cobalt(III)

trans   *cis*

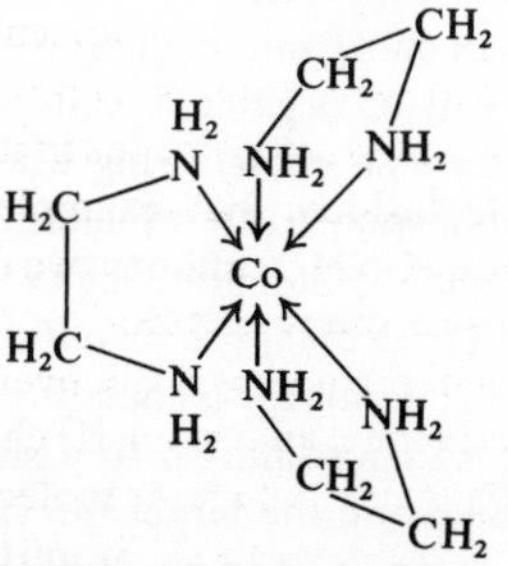

tris(ethylenediamine)cobalt(III)

Oxidation of cobalt(II) in the presence of excess ethylenediamine results in the formation of the complex ion $[Co\ en_3]^{3+}$. The coordination number of cobalt is again 6 and it has octahedral stereochemistry.

In the examples of ethylenediamine complexes so far mentioned, coordination of the metal establishes a heteroatomic ring. This process of ring formation is known as *chelation*. The word *chelate*, which describes the ring, was originally proposed in 1920 by Morgan and Drew. It is derived from the Greek *chele*, meaning a lobster's claw. Ligands like ethylenediamine which act in this way are referred to as *chelating agents* and the complexes they form as *metal chelates*. Chelation changes, often profoundly, the chemical and physical characteristics of the constituent metal ion and ligands and has far-reaching consequences in the realms of chemistry and biology.

In ethylenediamine, the two donor atoms are separated by an ethylene chain. Carbon is the element that has *par excellence* the capacity for chain-formation and so it is not surprising that many organic substances act as multidentate chelating agents. In addition to neutral molecules like ethylenediamine, many chelating agents function by losing one or more protons to produce anions which then coordinate to the metal atom. For example, glycine, NH_2CH_2COOH, loses its carboxyl-group proton and

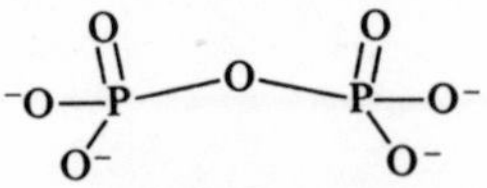

$$\text{pyrophosphate ion}$$

chelates as the glycinate ion via the nitrogen and one of the oxygen atoms. Oxalic acid, $(COOH)_2$, is another well-known chelating agent. This can lose two protons and then coordinate as the bidentate oxalate ion. Ethylenediamine, glycine, and oxalic acid are three of the simplest organic chelating agents and there are numerous more complex organic compounds which show similar properties.

A number of important inorganic chelating agents are also known, functioning in the same way as organic ligands. For example, the pyrophosphate ion, $P_2O_7^{4-}$, forms a chelate ring when a negatively charged oxygen attached to each phosphorus coordinates to the same metal atom.

Metal complexes containing bidentate ligands are usually easier to prepare and more difficult to decompose than those involving comparable unidentate molecules. In general, for a chelating ligand which contains z donor atoms all capable of coordination to a single metal ion, $(z - 1)$ chelate rings will be formed and the larger the value of z the greater the stability of the resulting complex. The enhanced stability associated with chelation is called *the chelate effect* and is considered in Chapter 4.

The value of z for a given ligand is indicated by the appropriate Latin prefix: bi-(2), tri-(3), quadri-(4) quinque-(5), sexa-(6), and so on. Thus, N-(2-aminoethyl)ethylenediamine contains three donor nitrogens and is therefore tridentate, nitrilotriacetic acid (NTA) is quadridentate, and ethylenediaminetetraacetic acid (EDTA) is sexadentate.

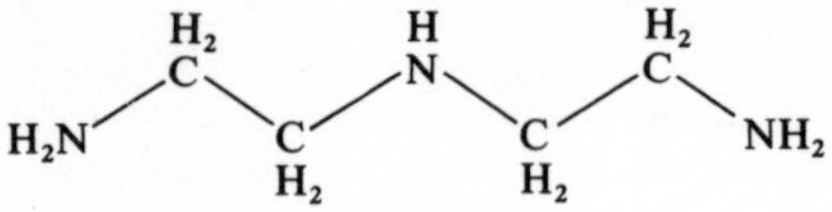

N-(2-aminoethyl)ethylenediamine

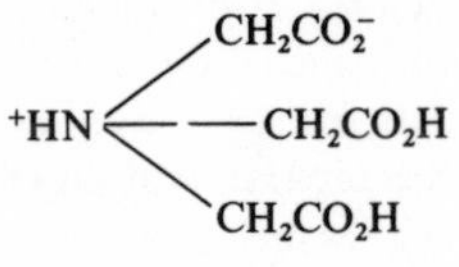

NTA

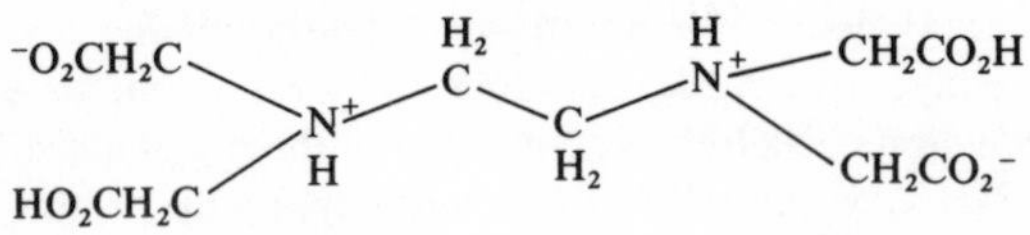

EDTA

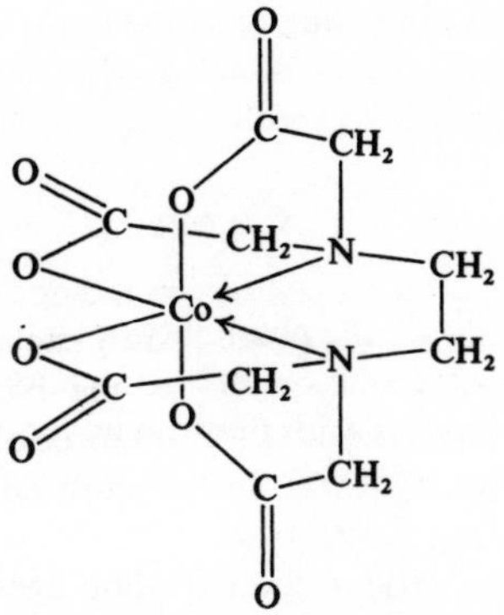

Cobalt (III) coordinated with EDTA

NTA and EDTA are particularly useful chelating agents. When all their donor atoms are coordinated to a metal, three and five chelate rings are formed respectively by NTA and EDTA. Both ligands accordingly form very stable complexes with metals. Moreover, the coordination number of six for metals like cobalt(III) is completely satisfied by coordination with one EDTA ligand.

Equilibria in solution between metal ions and chelating ligands

A chelating ligand is a special kind of complexing agent, and so the equilibria established in a solution containing this and a metal ion can be described in the same terms as those used for complex formation in general.

We are primarily concerned here with reactions in aqueous media. Each metal ion M^{n+} is hydrated† and its reaction with a ligand L involves replacement of water molecules in its coordination shell.

When L is an uncharged unidentate ligand, the formation of complexes proceeds in a stepwise manner with successive replacement of water molecules:

$$M(H_2O)_x^{n+} + L \rightleftharpoons M(H_2O)_{x-1}L^{n+} + H_2O \qquad (K_1)$$

$$M(H_2O)_{x-1}L^{n+} + L \rightleftharpoons M(H_2O)_{x-2}L_2^{n+} + H_2O \qquad (K_2)$$

and so on. K_1 and K_2 are equilibrium constants for the above reactions.

For a bidentate ligand (L-L), two water molecules are replaced in each step:

$$M(H_2O)_x^{n+} + (L-L) \rightleftharpoons M(H_2O)_{x-2}(L-L)^{n+} + 2H_2O.$$

†The solvation of metal ions is described by Pass in *Ions in solution* (3): *inorganic properties* (OCS 7).

In the special case of a multidentate ligand (L_x), for which x is the same as the coordination number of the metal, all coordinated water molecules may be replaced during reaction with one ligand molecule:

$$M(H_2O)_x{}^{n+} + (L_x) \rightleftharpoons M(L_x)^{n+} + xH_2O.$$

In solution, there is always the possibility of competing reactions which can affect the extent of chelation of metal ions. For example, since hydrated metal ions behave as acids they can undergo *hydrolysis* by releasing one or more protons from coordinated water molecules:

$$M(H_2O)_x{}^{n+} \rightleftharpoons M(H_2O)_{x-1}(OH)^{(n-1)+} + H^+(aq)$$

$$M(H_2O)_{x-1}(OH)^{(n-1)+} \rightleftharpoons M(H_2O)_{x-2}(OH)_2{}^{(n-2)+} + H^+(aq).$$

The extent of hydrolysis depends on a number of factors, including the polarizing power of the metal ion and the pH of the solution. It is clear from equations that the formation of hydroxo complexes can be repressed by the addition of acid. Although this may have the desired effect of ensuring that the metal ion is predominantly present as $M(H_2O)_x{}^{n+}$, it may promote competing reactions, namely *protonation*, which involve the ligand. By virtue of their donor capabilities, ligands have basic properties and so will tend to react with acids to give protonated species:

$$L + H_3O^+ \rightleftharpoons LH^+ + H_2O.$$

A bidentate ligand like ethylenediamine can take up one or two protons:

$$en + H_3O^+ \rightleftharpoons en\ H^+ + H_2O$$

$$en\ H^+ + H_3O^+ \rightleftharpoons en\ H_2{}^{2+} + H_2O$$

The amount of free (that is, non-protonated) ethylenediamine available for chelation with the metal ion, and hence the extent to which chelate formation occurs, varies accordingly with the pH and the magnitude of the equilibrium constants.

The same considerations apply to ligands that coordinate as anions. For example, oxalate ion exists in aqueous solution in equilibrium with its protonated forms:

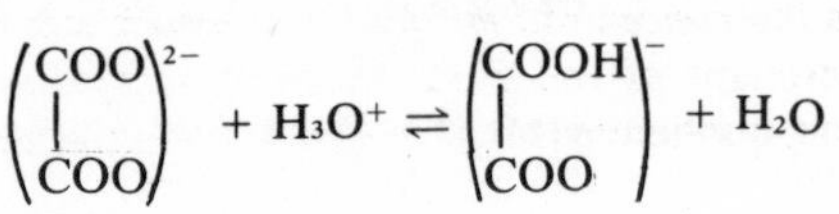

$$\begin{pmatrix} COO \\ | \\ COO \end{pmatrix}^{2-} + H_3O^+ \rightleftharpoons \begin{pmatrix} COOH \\ | \\ COO \end{pmatrix}^- + H_2O$$

$$\begin{pmatrix} COOH \\ | \\ COO \end{pmatrix} + H_3O^+ \rightleftharpoons \begin{pmatrix} COOH \\ | \\ COOH \end{pmatrix} + H_2O \; .$$

These protonation reactions therefore compete with the chelating action of oxalate towards metal ions.

Thermodynamic aspects

Each of the reactions given above is an equilibrium process and so an equilibrium constant may be defined in terms of the relative activities of the various species present. Thus for the reaction between a hydrated metal ion and a unidentate ligand, the equilibrium constant

$$K_1 = \frac{a(ML) \cdot a(H_2O)}{a(M) \cdot a(L)} \; .$$

Similarly,

$$K_2 = \frac{a(ML_2) \cdot a(H_2O)}{a(ML) \cdot a(L)} \; .$$

For brevity, coordinated water molecules and charges are omitted.

For reactions in dilute aqueous solution, $a(H_2O)$ remains virtually constant and so, for the formation of ML, we can define another constant, K_1T:

$$K_1T = \frac{K_1}{a(H_2O)} = \frac{a(ML)}{a(M) \cdot a(L)} = \frac{[ML] \cdot \gamma(ML)}{[M][L] \; \gamma(M) \cdot \gamma(L)} \; ,$$

where K_1T is the *thermodynamic stability (or formation) constant* of the complex ML, square brackets denote concentrations, and $\gamma(M)$, $\gamma(L)$, and $\gamma(ML)$ are activity coefficients. K_1T is a true thermodynamic constant and is independent of ionic strength.

Another constant, K_1C, can be defined in terms of the concentrations of the species present:

$$K_1C = \frac{[ML]}{[M][L]} \; .$$

K_1C is known as the *concentration stability constant* and its value varies with the ionic strength of the medium. With increasing dilution, the activity coefficient quotient $\gamma(ML)/\gamma(M) \cdot \gamma(L)$ approaches unity and so $K_1C \to K_1T$.

Concentration stability constants are useful in describing quantitatively systems for which activity-coefficient data are inadequate or unknown. When concentration constants have been measured over a range of ionic strength, extrapolation of the values to infinite dilution gives a value for $K_1{}^T$.

The thermodynamic constant $K_1{}^T$ is related to the standard free energy change ΔG^o of the reaction by†

$$\Delta G^o = -RT \ln K_1{}^T.$$

ΔG^o is made up of an enthalpy (ΔH^o) and an entropy (ΔS^o) term:

$$\Delta G^o = \Delta H^o - T\Delta S^o.$$

These two relations provide the starting-point for the consideration of the significance of enthalpy and entropy changes accompanying chelation (Chapter 4).

†See Smith's *Basic chemical thermodynamics* (OCS 8) for a discussion of this equation.

1. Properties of ligands and chelate rings

Conditions for chelation

For chelation to occur, the ligand must contain at least two donor atoms capable of bonding to the same metal atom. Elements that act as donors are the more electronegative ones on the right-hand side of the periodic table. They are primarily found in Group V (nitrogen, phosphorus, arsenic, and antimony) and in Group VI (oxygen, sulphur, selenium, and tellurium). Although numerous complexes of Group VII elements, the halogens, are known, the simple halide ions behave either as unidentate ligands in mononuclear complexes or, sometimes, as bridging atoms in polynuclear complexes. Covalently bound halogens do not act as donors in chelating systems.

Donor atoms may form part of a basic or an acidic functional group. A basic group is one that contains an atom carrying a lone pair of electrons which may interact with a metal ion (or a proton). Some of the more important are: $-NH_2$ (amino), $-NH$ (imino), $-N=$ (tertiary acyclic or heterocyclic nitrogen), $=O$ (carbonyl), $-O-$ (ether, ester), $=N-OH$ (oxime), $-OH$ (alcohol), $-S-$ (thioether), $-PR_2$ (substituted phosphine), and $-AsR_2$ (substituted arsine).

An acidic group loses a proton and coordinates with a metal atom. These groups include: $-CO_2H$ (carboxylic), $-SO_3H$ (sulphonic), $-PO(OH)_2$ (phosphoric), $-OH$ (enolic and phenolic), $=N-OH$ (oxime), and $-SH$ (thioenolic and thiophenolic).

Another condition for chelation is that the functional groups should be so located in the ligand that the formation of a ring including the metal atom is sterically possible. For example, in ethylenediamine the two amino groups are separated by two linked carbon atoms and this carbon bridge confers some degree of flexibility on the molecule so that it easily adopts conformations in which the nitrogen atoms occupy adjacent positions in the coordination shell of a metal ion without much departure from the 'natural' tetrahedral bond angles of four-covalent carbon. Thus chelation of ethylenediamine imposes relatively little strain on the ligand molecule. The chelating properties of propylenediamine (1,3-diaminopropane), $H_2NCH_2CH_2CH_2NH_2$, and longer-chain aliphatic diamines are similar to those of ethylenediamine because the flexibility of the carbon chain permits chelation to occur easily.

Even when the donor atoms appear to be appropriately situated for chelation, this may be rendered difficult or impossible by the presence of bulky groups substituted elsewhere in the molecule so that close approach of ligand to metal is hindered or prevented altogether. This type of effect is known as steric hindrance and some examples are given below.

The presence of functional groups and the fulfilment of steric requirements are necessary but not always sufficient conditions for chelation to occur: experimentally, this may still not be possible. For example, in aqueous solutions of low pH, competitive protonation of the ligand may be so extensive that some, if not all, of the donor atoms are unable to coordinate to the metal ion.

Colour in chelates

One of the most fascinating aspects of the chemistry of metal chelates is the almost infinite variety of colours displayed. In the coordination compounds of a single metal such as cobalt, there is an astonishing range of colours, each depending on the particular group of ligands present. This phenomenon of colour extends from low molecular weight complexes, like those studied originally by Werner, to chelates of very high molecular weight, such as those found in nature. Among the latter are chlorophyll, the green colouring matter in plants, and the blue haemocyanin and red haemoglobin pigments, responsible respectively for the transport of oxygen in the blood of molluscs and mammals.

Colour results from the absorption of radiation in some part of the visible region of the electromagnetic spectrum.† Absorption usually extends into the near infrared or ultraviolet regions and is associated with the excitation of electrons within the complex from lower to higher energy levels.

Three types of electronic transition giving rise to absorption spectra can be distinguished, although it is not always possible to define these separately in a particular complex. They are:

(a) *d-d transitions*. These occur between filled d-orbitals and empty or half-filled d-orbitals of the metal ion. The colours shown by transition-metal ions, both in solids and in solution, are commonly attributable to these transitions; the number and energy of the electronic transitions depending on the metal, its oxidation state, and the ligands with which it is combined. Electronic transitions between d-orbitals of the same quantum shell are, strictly speaking, forbidden by the Laporte rule. They are observed in practice because some loss in symmetry of the complex ions permits them to occur. The intensity of absorption is usually low.

(b) *Transitions within ligands*. Organic ligands which contain delocalized π-electron systems show characteristic absorption in the ultraviolet or in the shorter wavelength region of the visible spectrum. This is generally

†See Atkins's *Quanta — a handbook of concepts* (OCS 21) for extra information on colour and electronic transitions.

manifested by the appearance of intense absorption bands resulting from $\pi^* \leftarrow \pi$ transitions (excitation of electrons from a lower energy π-bonding orbital to a higher energy π-antibonding orbital). In the case of chelating ligands which contain delocalized π-electrons, their colours and those of their metal chelates are intensified if one or more chromophoric groups are introduced into the molecular structure. These groups include C=O (carbonyl), C=S (thioketo), $-CH=N-$ (azomethine), $-N=N-$ (azo), $-N=O$ (nitroso), and $-NO_2$ (nitro). They serve to extend the conjugated system so that absorption bands are displaced to longer wavelengths, that is, more into the visible region. The same effect is produced if the number of aromatic rings within the molecule is increased. Loss of a proton from an acidic functional group or protonation of a basic group alters the colour of the ligand. Similarly, chelation of a metal ion usually causes a pronounced colour change. The intensity of absorption together with the colour changes produced by coordination can often be exploited analytically by using such ligands for the colorimetric determination of very small amounts of metals.

(c) *Charge-transfer transitions.* As their name suggests, these are transitions in which transfer occurs of an electron from the ligand to the metal or in the reverse direction. Transfer in the first direction is effectively a reduction of the metal; transfer from metal to ligand is an oxidation process. Charge-transfer transitions give rise to strong absorption bands, whose intensities are comparable with those bands resulting from $\pi^* \leftarrow \pi$ transitions.

Although the above division into three types of electronic transitions is convenient, the boundaries between them are often not clear-cut in practice and absorption processes taking place within a metal chelate should really be considered as involving orbitals extending over the whole complex.

Classes of ligands

Ligands can be classified according to the donor atoms present and then further divided in terms of the particular types of functional groups. This provides a basis for consideration of the properties of ligands that are of major significance in chelation.

Bidentate ligands

There are three main classes of bidentate ligand containing respectively two basic groups, one acidic and one basic group, and two acidic groups. Ligands with two basic groups coordinate as the neutral molecule and the resulting chelate has the same charge as that originally on the metal ion. Organic diamines fall into this category. If the ligand, for example, an amino carboxylic acid, contains one acidic and one basic group, coordination of a metal usually involves loss of the ionizable proton. Hence the charge on the metal ion is reduced by one unit for each ligand chelated. When chelation continues with the attachment of several ligands, uncharged or even anionic complexes can be formed. Some ligands of this type can, in certain experimental conditions, alternatively chelate as neutral molecules and then the charge on the metal ion is unaltered. When two acidic groups are present, for example, in a dicarboxylic acid, each ligand anion coordinated reduces the positive charge on the metal by two units. With such ligands, anionic complexes are usually formed whenever the metal reacts with excess ligand.

Ligands with two basic groups. Pre-eminent among these are ligands containing two nitrogen donors. They range from aliphatic acyclic compounds like ethylenediamine, to heterocyclic bases like 2,2′-bipyridyl and 1,10-phenanthroline.

Ethylenediamine itself has a special place in the historical development of coordination chemistry. It was one of the first chelating ligands used by Werner, and its complexes with cobalt and platinum provided much of the experimental data on which his theories of isomerism in coordination compounds were based.

The stability constants of mono(ethylenediamine) complexes of some transition metals of the first series are listed in Table 1.1. The magnitude of these values shows that considerable stability is associated with the five-membered chelate ring, particularly for metal ions near the end of the series.

Homologues of ethylenediamine, such as propylenediamine, have similar chelating properties. When the number of carbon atoms in the chain is increased, the size of the chelate ring becomes greater. Table 1.1 also includes data for lg K_1 for propylenediamine chelates. The values illustrate how an increase in ring size from five to six atoms results in a decrease in stability. It is generally found to be the case, at least as far as aliphatic ligands are concerned, that five-membered chelate rings have a greater stability than those with six or more atoms.

TABLE 1.1
Stability constants (lg K_1) for ethylenediamine and propylenediamine complexes of transition metals†

Ligand	Metal					
	Mn^{2+}	Fe^{2+}	Co^{2+}	Ni^{2+}	Cu^{2+}	Zn^{2+}
Ethylenediamine	2.75	4.32	5.94	7.51	10.72	5.79
Propylenediamine				6.39	9.98	

†Measured at 25 °C in an aqueous medium 1 M with respect to KCl or KNO₃.

TABLE 1.2

*Stability constants of ethylenediamine and
N-alkylethylenediamine complexes of copper and nickel†*

Complex	Ligand (L-L)			
	en	N-(CH₃)en	N-(C₂H₅)en	N-(n-C₃H₇)en
$Cu(L-L)_2^{2+}$	20.13	19.11	18.57	18.14
$Ni(L-L)_3^{2+}$	19.11	15.11	14.08	13.76

†Measured at 25 °C in an aqueous medium 0.5 M with respect
to KNO_3.

The effect on stability of the inclusion of bulky alkyl groups in the molecular framework of the ligand is illustrated by the stability constants of *N*-substituted ethylenediamine complexes. Table 1.2 gives the values for 1:2 complexes of Cu(II) and 1:3 complexes of Ni(II) with some *N*-alkylethylenediamines, $RNH_2CH_2CH_2NH_2$, where R = CH_3, C_2H_5, or n-C_3H_7. The basic strengths of these ligands and ethylenediamine are very similar and so they have approximately the same donor ability towards a hydrogen ion. However, as alkyl groups of increasing size are substituted in one of the amino groups, the stability of the copper or nickel chelate decreases. Thus, although the introduction of bulky groups does not affect the ability of the ligand to protonate, this clearly hinders its interaction with metal ions, an effect manifested by lower stabilities.

The two bases 2,2'-bipyridyl (bipy) and 1,10-phenanthroline contain the structural unit $-N=C-C=N-$. The two nitrogens are located in heteroaromatic rings and so their environment is very different from that of the donor atoms in ethylenediamine. The stabilities of mono complexes with some transition metal ions (Table 1.3) are, with the exception of the copper complex, higher than those of the corresponding complexes of ethylenediamine. The phenanthroline structure therefore appears to be particularly favourable for chelation.

TABLE 1.3

*Stability constants (lg K_1)
for 1,10-phenanthroline complexes
of transition metals†*

Mn^{2+}	Co^{2+}	Ni^{2+}	Cu^{2+}	Zn^{2+}
4.13	7.25	8.8	9.25	6.55

†Measured at 20 °C in an aqueous
solution 0.1 M with respect to alkali-
metal nitrate.

2,2'-bipyridyl   1,10-phenanthroline

Increased stability may be related to the heteroaromatic nature of the phenanthroline ligand. This contains low-lying vacant π-orbitals which can interact with filled d-orbitals of π-symmetry on the metal ion. Thus, in addition to the donation of electron density from ligand to metal, there is also a transfer of charge, in the case of those metals with appropriately filled d-orbitals, from metal to ligand. This 'two-way' electron transfer enhances the strength of the metal-ligand bonds.

The ability of bipyridyl and phenanthroline to receive electrons from the metal is believed to be responsible for the effectiveness of these ligands in stabilizing low oxidation states of some transition metals, for example, vanadium(0) in V(bipy)$_3$ and chromium(+1) in Cr(bipy)$_3^+$.

Analytical applications of these two ligands are important because some of their metal complexes have very intense colours. For example, absorbances of the copper(I) and iron(II) chelates are so great that microgram quantities of the metals can be readily detected. Not all transition metals give strongly coloured complexes. For instance, the manganese(II) chelates are almost colourless. The ligands can therefore be used as selective colorimetric reagents for copper and iron in the presence of other metals which do not give coloured chelates.

The phenomenon of steric hindrance has been extensively studied in substituted derivatives of bipyridyl and phenanthroline. Thus the substitution of methyl groups in the 6- and 6'-positions of bipyridyl or in the 2- and 9-positions of phenanthroline inhibits the formation of the usual intensively coloured tris-chelates with iron(II) because the methyl groups in one ligand sterically interfere with other ligands attached to the same metal ion. These dimethyl-substituted ligands are still able to form highly coloured bis-chelates with copper(I). Hence they can be used as specific analytical reagents for copper, even when other metals, including iron, are present.

Elements below nitrogen in Group V also show donor properties but these are less marked than those of nitrogen-containing ligands. The ability to act as donor falls off as the atomic number increases and the stabilities of complexes between a metal and a substituted phosphine and its arsenic and antimony analogues follow the sequence: P > As > Sb.

(R′ = C$_6$H$_5$ or C$_2$H$_5$)
substituted ethylene

(R = CH$_3$, C$_2$H$_5$, or C$_6$H$_5$)
o-phenylenebis(dialkyl- or
diarylphosphine)

Compounds of these heavier Group V elements are more effective ligands if they contain at least two donor atoms situated so that chelation can occur. The increase in stability associated with ring-formation makes up, in part at least, for the relatively poor donor capability.

Many diphosphines have been examined as ligands. Derivatives of diphosphine itself, H$_2$PPH$_2$, contain two donor atoms linked by a single bond and cannot, on geometrical grounds, undergo chelation. As in the case of organic diamines, chelation first appears to be possible when the donor atoms are separated by an ethylene bridge. This is so in the substituted diphosphines, R$_2$PCH$_2$CH$_2$PR$_2$, for example, in diphenylphosphino-ethane, where R = C$_6$H$_5$.

Phosphorus, arsenic, and antimony all have vacant d-orbitals which can accept electrons from coordinated metal atoms and for this reason substituted phosphines, arsines, and stibines can function as π-electron acceptors as well as σ-electron donors. The tendency of metals to donate electrons into d-orbitals of phosphines and other Group VB donors is expected to be greater in zero-valent metal complexes than in those of higher oxidation states. Such π-bonding is therefore generally very important for the stabilization of low oxidation states.

In other ligands, the bridging carbon atoms are linked by a bond of multiple character, as in disubstituted ethylenes and in *o*-phenylenebis-(dialkyl- or diarylphosphines).

Arsenic analogues of some of the foregoing phosphorus ligands have also been synthesized. One of the most widely investigated of these is *o*-phenylenebis(dimethylarsine), noteworthy because of its ability to stabilize unusual oxidation states and coordination numbers of the transition metals.

o-phenylenebis(dimethylarsine)

Ligands with one acidic and one basic group. There are many ways in which an acidic and a basic group can be arranged within an organic molecule so that chelation is possible, and consequently a great diversity of such ligands is known.

One reaction commonly observed with these ligands is the formation of electrically neutral chelates. This happens when the process of chelation simultaneously satisfies the oxidation number and coordination number of the metal ion. For example, a divalent metal ion M^{2+}, of coordination number 4, reacts with a bidentate monobasic ligand HR according to:

$$M^{2+} + 2HR \rightleftharpoons MR_2 + 2H^+.$$

Similarly, a trivalent ion M^{3+}, of coordination number 6, chelates with HR:

$$M^{3+} + 3HR \rightleftharpoons MR_3 + 3H^+.$$

In general, whenever the coordination number is twice the oxidation number, chelation of this type of ligand can lead to the formation of an uncharged complex. When a metal ion is coordinated in this way, the properties of the complex are very different from those of simple salts of the metal. The uncharged metal chelate often shows general resemblances to organic compounds in its physical properties. This is hardly surprising, for the immediate environment of the metal ion is the surrounding shell of hydrophobic organic ligands. For this reason, the chelate is generally much more soluble in organic solvents than in aqueous solutions.

Complexes produced by the reaction between a metal ion and a molecule carrying acidic and basic functional groups are commonly designated as 'inner complexes'. This term was originally applied only to uncharged complexes formed in this way but now more generally to cationic, uncharged, or anionic chelates formed in this way.

The $1,3(\beta)$-keto-enols are a well-known class of organic chelating agents containing one acidic and one basic group. A β-keto-enol exists in tautomeric equilibrium with the β-diketo form:

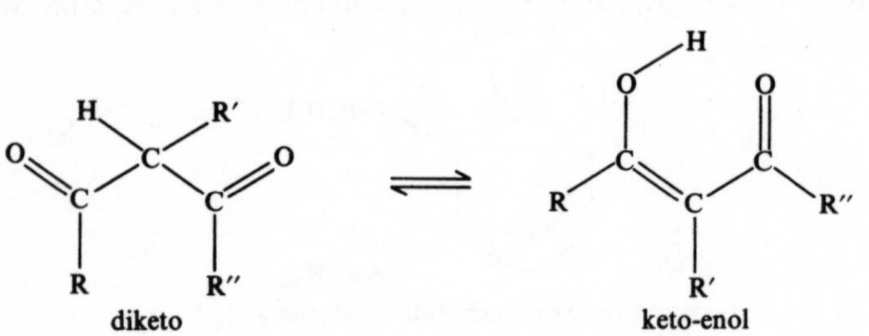

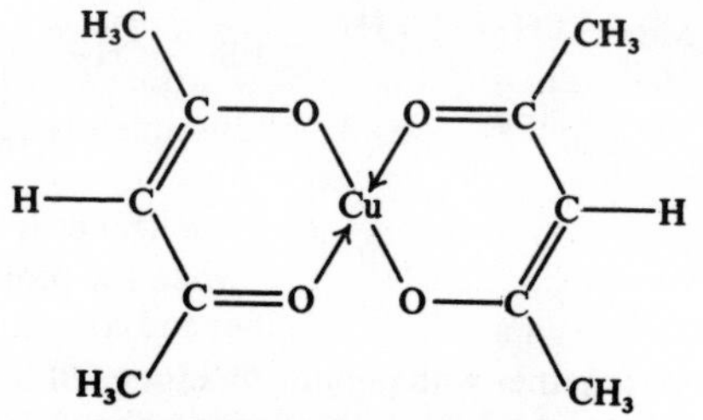

bis (acetylacetonato) copper (II)

The enolic hydrogen atom is intramolecularly bonded to the ketonic oxygen. The enol group is acidic and its proton is replaceable by metals. Simultaneous coordination of the keto-group oxygen completes the formation of a six-membered ring. Two such rings are found, for example, in bis(acetylacetonato)copper(II).

Thio-derivatives of β-diketones have similar chelating properties. These exist largely in the thiol form. Like its oxygen analogue, this tautomer contains an intramolecular hydrogen bond.

Chelation involving two oxygen-containing functional groups is possible in aromatic ligands where the groups are situated in the *ortho*-position to one another. For example, salicylaldehyde contains a weakly acidic OH group which can lose its proton and coordinate as a negatively charged ion. Chelation (producing a six-membered ring) is completed by simultaneous coordination through the carbonyl oxygen of the aldehyde group. Chelation is sterically impossible wherever two such groups are substituted in *meta*- or *para*-positions in the benzene ring.

α-Amino acids are an important class of bidentate ligand. The simplest ligand of this kind is glycine, NH_2CH_2COOH. In aqueous solution between pH = 4 and pH = 9, this exists as the zwitterion, $H_3^+NCH_2CO_2^-$; this loses a proton from the $-NH_3^-$ group and chelates a metal ion to form a five-membered ring. This acid and other amino acids are components of protein molecules which consist essentially of α-amino acids linked together by peptide ($-CONH-$) linkages. Terminal $-NH_2$ and $-CO_2H$

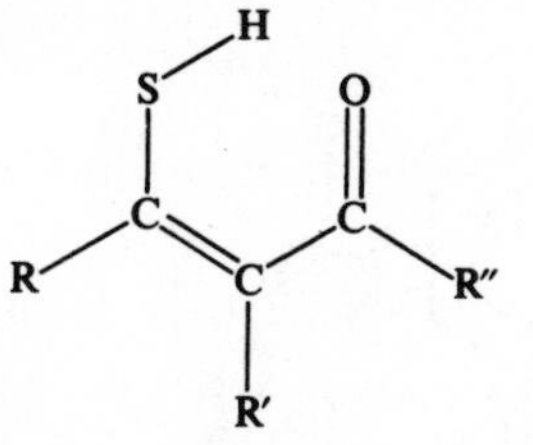

β-keto-thiol

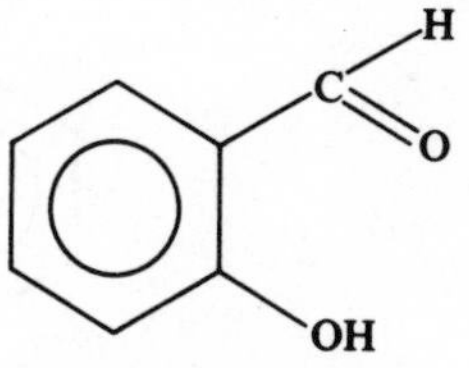

salicylaldehyde

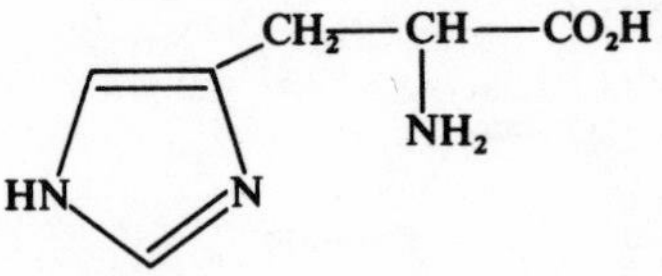

histidine 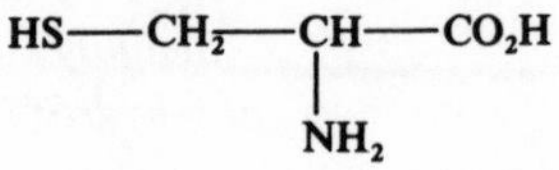   cysteine

groups are present, together with peptide linkages and side-chains. These are all potential sites for the coordination of metals. Two amino acid components of proteins with a particular affinity for metal ions are histidine and cysteine. As well as $-NH_2$ and $-CO_2H$ groups histidine has an imidazole ring and cysteine a thiol group, both of which can participate in bonding to a metal. In some structures, known as metalloproteins, the metal ion and protein are very firmly linked together so that the metal is an integral part of the molecular structure. Complexes of this type play vital roles in many biological processes.

Another well-known chelating agent containing an acidic $-OH$ group and a basic nitrogen atom is 8-hydroxyquinoline (8-quinolinol, HOx). As would be expected from its constitution, the complexing ability of this ligand is strongly pH-dependent. In solutions of high pH, the anionic ligand Ox^- predominates. The uncharged molecule is only slightly soluble in water but in acidic solutions the solubility increases again, because protonation of the ligand to give the cationic species H_2Ox^+ takes place. pK values for the dissociation of the first and second protons from H_2Ox^+ are respectively pK_1 5.02 and pK_2 9.81.

8-Hydroxyquinoline is used extensively in the analysis of metals by gravimetric, volumetric, or colorimetric procedures. Its properties can be modified by the introduction of substituents in the heterocyclic or benzene ring. For example, the insertion of a sulphonic acid grouping results in much greater water solubility. The substitution of halogens in the molecule increases the acidity of the protonated forms. Thus, for 5,7-dichloro-8-hydroxyquinoline in water, $pK_1 = 2.9$ and $pK_2 = 7.5$, and consequently the metal complexes of this ligand are formed and can be extracted at lower pH values than those of 8-hydroxyquinoline itself.

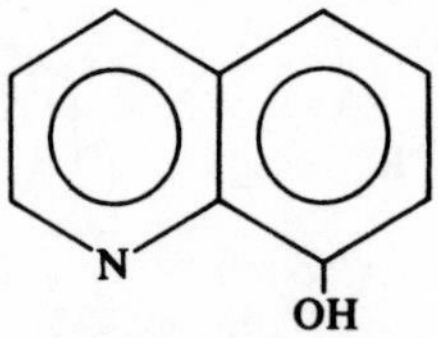

8-hydroxyquinoline

TABLE 1.4

Stability constants (lg K_1K_2)
of 8-hydroxyquinoline complexes

Ligand	Metal				
	Mn^{2+}	Co^{2+}	Ni^{2+}	Cu^{2+}	Zn^{2+}
8-Hydroxyquinoline	15.5	19.7	21.4	26.2	18.9
2-Methyl-8-hydroxyquinoline	14.0	18.5	17.8	23.8	18.7
4-Methyl-8-hydroxyquinoline	15.5	20.0	22.3	26	20.2

The effect of steric hindrance in substituted derivatives is illustrated by the lower stability of 2-methyl-8-hydroxyquinoline complexes compared with those of 8-hydroxyquinoline or 4-methyl-8-hydroxyquinoline (Table 1.4). The acid strengths of the three ligands are approximately constant but for the transition metal ions Mn^{2+} to Zn^{2+} inclusive, the values of lg K_1K_2 are smallest, in all cases, for the complexes of the 2-methyl-substituted ligand. The presence of the methyl group adjacent to the nitrogen donor reduces the accessibility of the latter to metal ions, so leading to a decrease in stability of the complexes.

The α,β-dioximes are another group of chelating ligands which have been widely studied. These molecules contain two oxime groups substituted on adjacent carbon atoms and form inner bis-complexes with metal ions such as Ni(II), Pd(II), and Cu(II). Each ligand loses on coordination one of its oxime-group protons, and the anionic form bonds to the metal via the two nitrogen atoms. Five-membered rings of considerable stability are established. Of the metal complexes, bis(dimethylglyoximato)nickel(II) is probably the best-known. This has a characteristic deep pink colour which is exploited in colorimetric procedures for the determination of trace amounts of nickel. The complex also has very low solubility in water and the precipitation of this from aqueous solution is the basis of a gravimetric method for the analysis of nickel. The bis(dimethylglyoximato)cobalt(II) complex, or 'cobaloxime', is of special interest because it resembles chemically and structurally the biologically important cobalt complex, vitamin B_{12}.

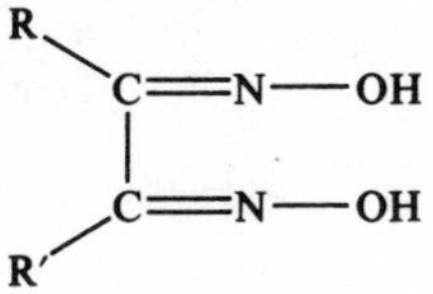

α,β-dioxime

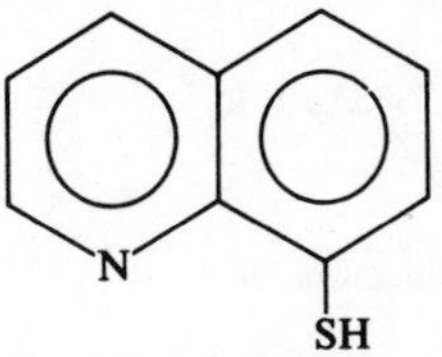

8-mercaptoquinoline

Several types of bidentate ligand are known in which the coordinating groups are a thiol group and a basic nitrogen atom. These include 2-amino-ethanethiol, $HSCH_2CH_2NH_2$, and the sulphur analogue of 8-hydroxy-quinoline, 8-mercaptoquinoline.

Thiosemicarbazide is another ligand which exists in tautomeric forms. The thio-keto form can act as a neutral ligand and the thio-enol form, by loss of a proton, as a negatively charged chelating ion.

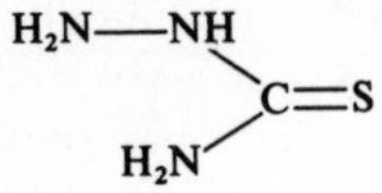

thio-keto 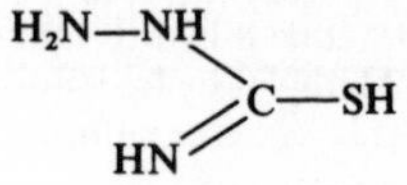   thio-enol

Similarly, 3-mercapto-1,5-diphenylformazan (dithizone) exhibits tautomerism between a thio-keto and a thio-enol form. In the thio-enol form, the thiol proton is replaceable by metal ions. Chelation is possible via the sulphur atom and one of the nitrogens. Dithizone itself is a violet-black solid, soluble in many organic solvents to give deep-green solutions. The metal complexes (dithizonates) are variously coloured violet, red, orange, or yellow in organic solvents, depending on the metal involved. The colours are very intense and, for many heavy metals, serve as a sensitive test for the presence of trace amounts.

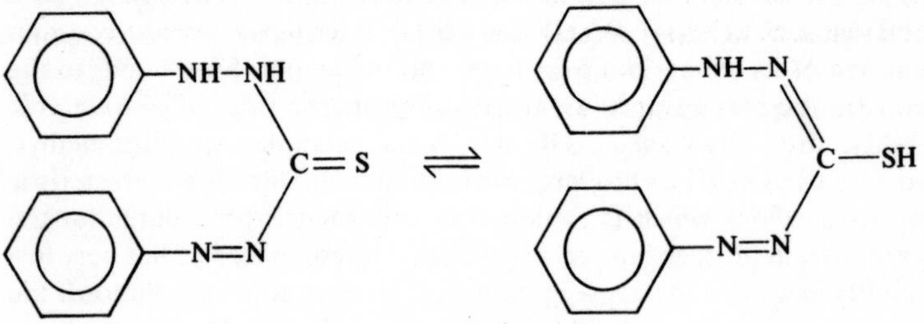

tautomeric forms of dithizone

Ligands with two sulphur atoms capable of coordination to a metal include the dithiocarbamates, xanthates, and dithiophosphates.

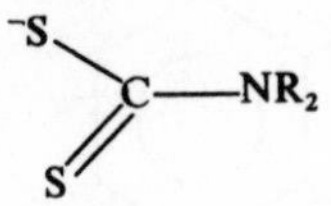

dithiocarbamate 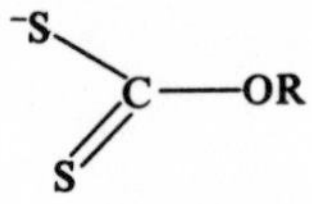   xanthate 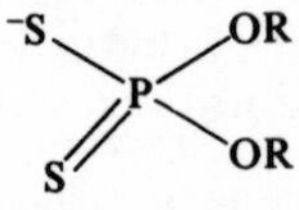   dithiophosphate

R is an alkyl group

Chelation of these ligands occurs via the two sulphur atoms with the establishment of a four-membered ring. Normally, a high degree of strain and hence instability is associated with small chelate rings but here electron delocalization within the rings which extends into the rest of the ligand exerts a counterbalancing stabilizing effect.

Ligands with two acidic groups. These include a number of simple inorganic acids as well as many organic ligands.

The chelating behaviour of simple inorganic anions containing three or four oxygens is well established. Examples include CO_3^{2-}, SO_4^{2-}, PO_4^{3-}, and CrO_4^{2-}. For example, the complex cation $[Co(NH_3)_4CO_3]^+$ contains a bidentate carbonate ion, the cobalt(III) ion achieving its customary coordination number of 6.

The sulphate ion shows a similar bidentate function in $Co(en)_2SO_4$ Br and $Pd(phen)SO_4$ (phen = 1,10 phenanthroline).† These anions can alternatively coordinate as a unidentate ligand or even as bridging groups in polynuclear complexes, where two of the oxygens coordinate to different metal atoms.

We may note that perchlorate and nitrate ions can also act in some complexes as chelating agents. These are, of course, derived from monobasic acids but delocalization of negative charge within the anion ensures that the two oxygens involved in chelation are exactly equivalent to one another and so it is not surprising that perchlorate behaves stereochemically in a similar manner to sulphate, and correspondingly nitrate to carbonate. Bidentate perchlorate groups are found in, for example, $Ni(CH_3CN)_2(ClO_4)_2$, and bidentate nitrato groups are present in some anhydrous metal nitrates, for example, $Cu(NO_3)_2$.

Copper nitrate

The simplest organic dibasic acid is oxalic, and the planar oxalate ion $C_2O_4^{2-}$ behaves as a bidentate ligand, forming a five-membered ring.

When we consider the chelating property of the homologous series of dibasic acids beyond oxalic, the stability of the complex with a particular metal is found to decrease progressively as the number of methylene groups in the acid increases, that is, the stability constants of the metal complexes

†Some metal complexes of oxo-anions existing in aqueous solution are of the outer-sphere type. There is no direct bonding linking the anion to the metal but bonding is established via the water molecules in the hydration shell of the metal ion. In some cases, for instance, sulphato complexes of first-row transition metals, both outer- and inner-sphere complexes can exist in equilibrium in aqueous solution.

follow the sequence:

oxalate	malonate	>	succinate	>	glutarate
$^-OOC\text{-}COO^-$	$^-OOC\text{-}CH_2\text{-}COO^-$		$^-OOC\text{-}(CH_2)_2COO^-$		$^-OOC\text{-}(CH_2)_3\text{-}COO^-$

Provided all these are indeed functioning as chelating ligands, then the ring becomes progressively less stable the greater, beyond five, the number of atoms present in the ring.

Aromatic ligands with two acidic groups are of importance in analysis. Examples of where chelation is through two oxygen atoms are 1,2-dihydroxy-3,5-disulphonic acid (Tiron), salicylic acid (*o*-hydroxybenzoic acid) and catechol (1,2-dihydroxybenzene). These react preferentially with ions such as Fe(III) and Ti(IV) which have a high affinity towards oxygen.

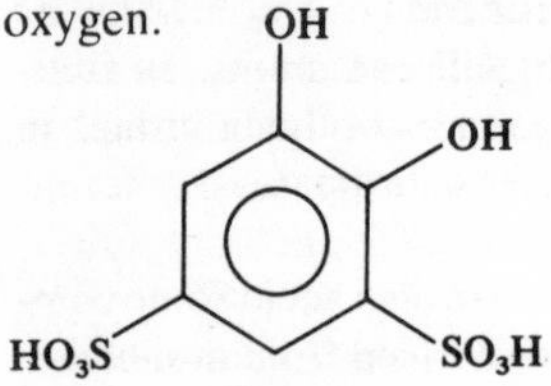

Tiron 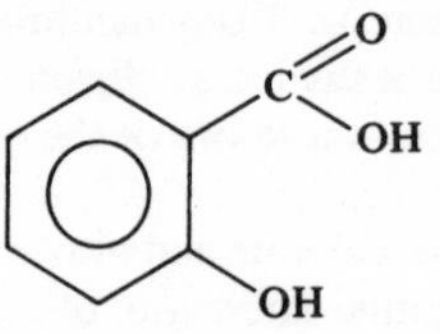   salicylic acid 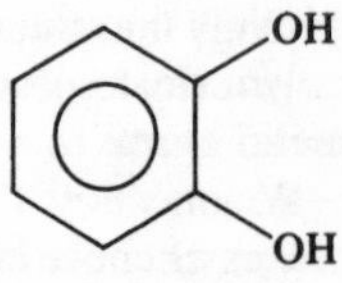   catechol

Sulphur-containing bidentate ligands include those based respectively on ethane and benzene. One of the best-known of the second type is toluene 3,4-dithiol (R′ = CH₃, R″ = H), a useful complexing agent applied in analysis.

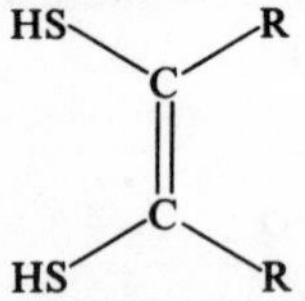

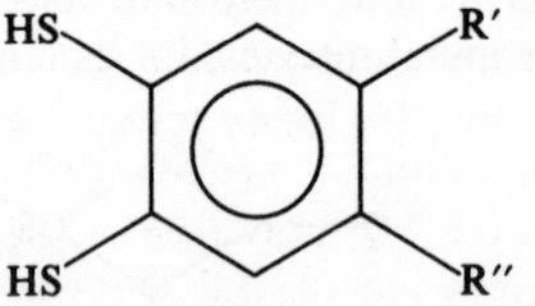

Ligands of this kind in which two sulphur atoms are linked to adjacent carbon atoms appear to stabilize some less common stereochemistries, for example, square-planar geometry in the bis-complexes and a trigonal prismatic arrangement of ligand atoms in tris-complexes of certain transition metals (Chapter 2).

Ligands of greater multidentate character

Many chelating molecules are known in which more than two donor atoms are present. In the great majority of cases, the donors concerned are oxygen or nitrogen. With increasing multidentate character, the ligand is able to occupy a larger number of positions in the coordination shell of the metal ion and, in the limiting case, all coordination positions can be filled by the donor atoms from one ligand molecule.

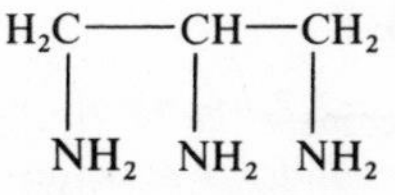

H$_2$C——CH—CH$_2$ H$_2$C——CH—CH—CH$_2$
| | | | | | |
NH$_2$ NH$_2$ NH$_2$ NH$_2$ NH$_2$ NH$_2$ NH$_2$

1,2,3-triaminopropane 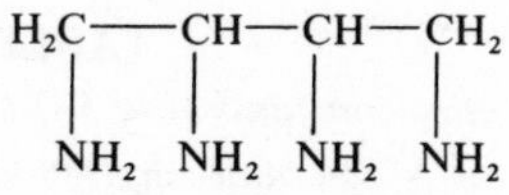   1,2,3,4-tetraaminobutane

In order to appreciate the most suitable design for multidentate ligands, we consider again the basic structural feature of the simple bidentate ligand, ethylenediamine. In this molecule, the donor nitrogens are attached to adjacent carbon atoms and, when chelation to a metal takes place, a five-membered ring is formed. One way of designing multidentate ligands is to use longer carbon chains and to continue with substitution of further primary amine groups on three, four, or more adjacent carbon atoms. Thus we might expect 1,2,3-triaminopropane to act as a tridentate, 1,2,3,4-tetraaminobutane as a quadridentate ligand and so on. In fact, although the tridentate behaviour of triaminopropane is well established, it is sterically impossible for the donor atoms in a molecule like 1,2,3,4-tetra-aminobutane to occupy four positions in the coordination shell of a single metal ion.

A much more favourable situation for multidentate action is achieved if one or more of the donors also functions as a linking atom within the molecule, thereby facilitating the introduction of a side-chain carrying another donor atom. This is the case, for example, in N-(2-aminoethyl)-ethylenediamine and N,N'-di-(2-aminoethyl)ethylenediamine. Either of these molecules is flexible enough to change to the appropriate conformation in which it can 'wrap around' the metal ion in such a way that all the nitrogens enter its coordinate shell without difficulty.

Comparison of stability-constant data (Table 1.5) for 1:1 complexes of divalent ions with 1,2,3-triaminopropane and N-(2-aminoethyl)ethylene-diamine shows clearly how the molecular structure of the ligand affects the stabilities of its complexes. Although in each case three donor nitrogens are involved in the formation of two five-membered chelate rings, the complexes are much less stable in the case of 1,2,3-triaminopropane

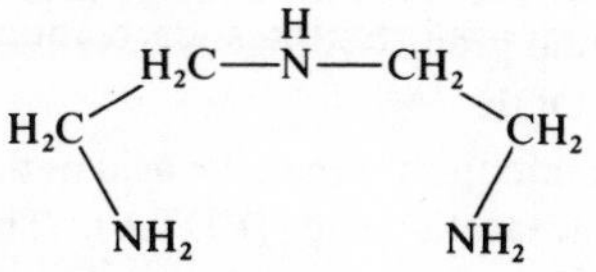

N-(2-aminoethyl) ethylenediamine

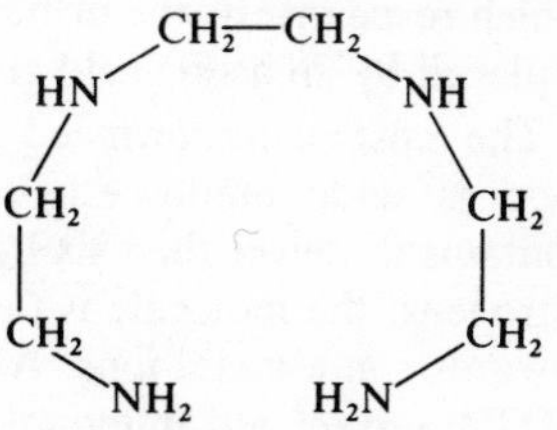

N,N'-di-(2-aminoethyl)
ethylenediamine

TABLE 1.5

*Stability constants (log K_1) for 1,2,3-triaminopropane
and N-(2-aminoethyl)ethylenediamine complexes
of transition metals†*

Ligand	Metal			
	Co^{2+}	Ni^{2+}	Cu^{2+}	Zn^{2+}
1,2,3-triaminopropane	6.8	9.3	11.1	6.75
N-(2-aminoethyl)ethylenediamine	8.1	10.7	16.0	8.9

†Measured at 20 °C in an aqueous medium 0.1 M in KCl.

(where considerable strain is developed by the ligand being forced to adopt the conformation necessary for tridentate coordination) than they are for *N*-(2-aminoethyl)ethylenediamine (for which there is virtually no strain within the ligand molecule).

Complexes of *N,N'* di-(2-aminoethyl)ethylenediamine are very stable and it is quite feasible to synthesize quinque- and sexadentate ligands of similar structure and hence metal complexes of even greater stability. There are, however, two practical disadvantages to the extensive use of such aliphatic polyamine ligands. First, the introduction of several nitrogen atoms into one molecule increases the basicity of the ligand. When it is used to complex metal ions in aqueous media, there is an increasing tendency for competitive protonation to occur; the pH of the solution increases; and there is a greater danger of the formation of hydroxo complexes, or even of precipitation of the metal hydroxide. Second, the number of metals that form stable complexes with multidentate nitrogen-containing ligands is not large compared with the number that combine with oxygen-containing ligands.

These disadvantages are overcome if basic nitrogens and acidic functional groups containing oxygen are incorporated in the same molecule, as in, for example, imidodiacetic acid $NH(CH_2COOH)_2$, or nitrilotriacetic acid $N(CH_2COOH)_3$. These are derivatives of glycine in which respectively one or both of the amino group hydrogens have been replaced by an acetic acid group $-CH_2COOH$.

The most well-known and widely used aminopolycarboxylic acid of this type is undoubtedly ethylenediaminetetraacetic acid (EDTA). This contains no fewer than six ligand atoms. Due to the presence of the two nitrogens, the molecule is flexible enough to act as a sexadentate ligand towards some metal ions. Aspects of the chemistry of the complexes of EDTA and related aminopolycarboxylic acids are described in Chapter 5.

The most important multidentate ligands that contain only nitrogen donors are those in which at least some of the nitrogen atoms form part of heterocyclic systems, for example, 2,2',6',2"-terpydridyl. This may be regarded formally as derived from 2,2'-bipyridyl by the replacement of a hydrogen atom in the 6'-position by a pyridyl group so that the ligand contains the following arrangement of atoms: $-N=C-C=N-C-C=N$. This

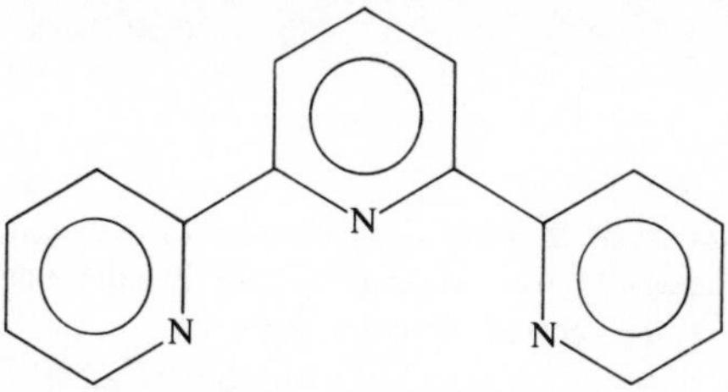

2,2',6',2"-terpyridyl

coordinates as a planar tridentate ligand with a metal to form two five-membered chelate rings. There are, of course, other ways in which three donor nitrogens can be linked to form a ligand which functions in a similar manner. Examples are 2(α-pyridylmethyleneaminomethyl)pyridine and pyridine-2-aldehyde 2'-pyridylhydrazone. In both molecules, the central donor nitrogen, unlike that in terpyridyl, is of a non-heterocyclic nature.

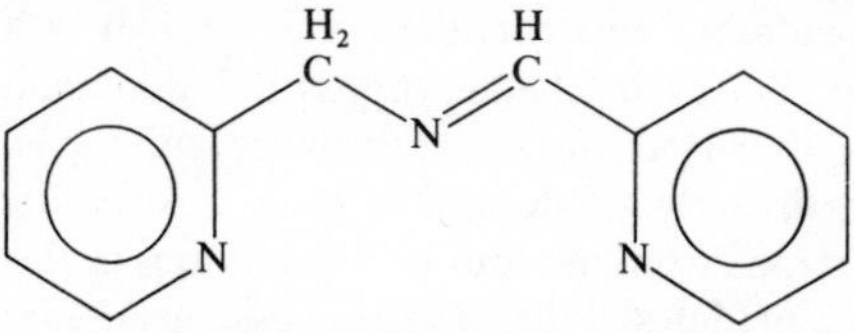

2(α-pyridylmethyleneaminomethyl)pyridine

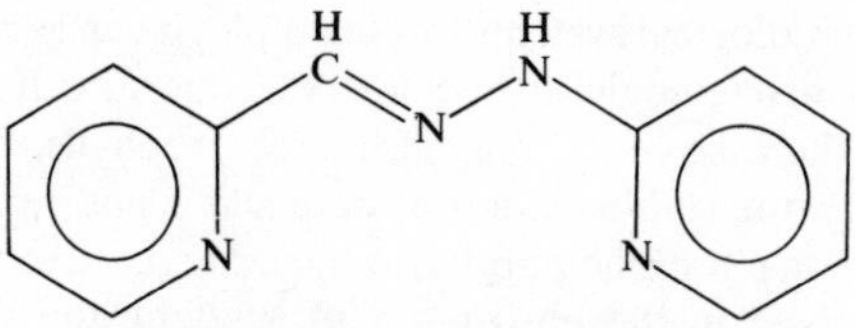

pyridine-2-aldehyde 2'-pyridylhydrazone

When we come to consider multidentate ligands with more than three donor nitrogens, we find that quadridentate ligands containing four pyrrole-ring nitrogens in a planar arrangement are chelating structures of special importance. These are exemplified by the corrin and porphin molecules.

corrin   porphin

The NH proton in corrin is lost upon coordination to a metal. The corrin structure is the basic chelating arrangement in the cobalt complex, vitamin B_{12}, and in the related cobamide coenzymes.

Porphin is the parent compound of a number of substituted derivatives known as porphyrins, which are important constituents of biologically important complexes. Loss of two protons from a porphyrin occurs on chelation and, with a divalent metal ion, this forms an uncharged inner complex. However, naturally occurring porphyrins contain side-chains carrying, for example, ionizable carboxyl groups and so the metal complexes are normally found to be charged.

The metal-porphyrin complexes are essential to the life of animal and plant organisms. For example, iron porphyrins are found in haemoglobins, myoglobins, peroxidases, and catalases. Chlorophyll is a magnesium-porphyrin complex. The cyclic planar porphyrin ring contains a conjugated pathway involving 18 π-electrons. In accordance with the Hückel rule,[†] it is found to be particularly stable and to show aromatic character. The donation of σ-electrons from the four nitrogen atoms to the central metal ion gives a complex of high stability. This process considerably modifies the redox potential of the chelated transition-metal ion tending to stabilize high formal oxidation states. The coordinated metal is able to function as a reactive centre in biological systems, for example, it can reversibly add on molecular oxygen in haemoglobin or can serve to activate oxygen in oxidizing enzymes. In the case of the magnesium chlorophylls, the metal concerned shows only one stable oxidation state and is not involved in redox reactions, but the macrocyclic porphyrin ligand is reactive in redox reactions, being involved in the absorption of sunlight and the process of photosynthesis.

Metal chelates derived from Schiff bases have been known for well over 100 years. These bases, of general formula RC=NR', can be synthesized by the condensation of primary amine with an active carbonyl group. Such compounds will be effective chelating agents whenever, in addition to the nitrogen atom, they possess at least one functional group so situated that, on complexation with a metal ion, a five- or six-membered ring is formed.

†See Atkins: *Quanta — a handbook of concepts* (OCS 21).

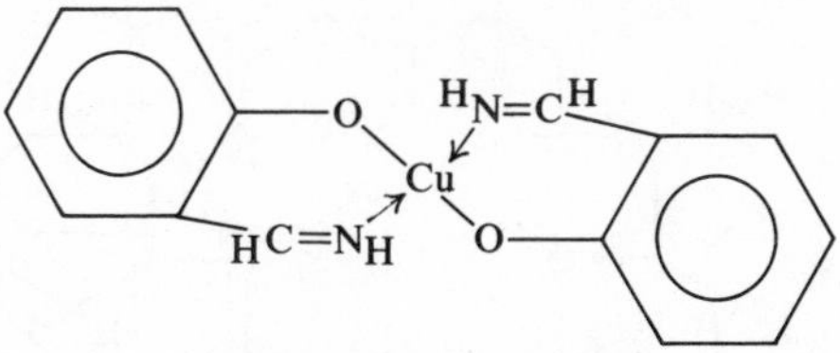

bis(salicylaldimino) copper(II)

One of the first chelates of a Schiff's base to be isolated was bis(salicyl-aldimino)copper(II), prepared by the reaction of copper(II) acetate with salicylaldehyde and aqueous ammonia. Schiff made the synthetically important discovery that chelates of these bases can be conveniently prepared by condensation of the appropriate amine with a pre-formed metal complex of the compound containing the carbonyl group. For instance, when a metal complex of salicylaldehyde is reacted with ethyl-enediamine, the product is a salicylaldimine chelate. In this case, the ligand is coordinating in a planar quadridentate manner.

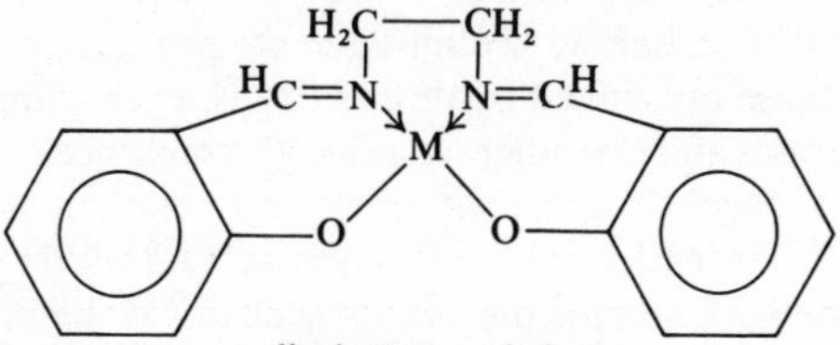

salicyladimine chelate

A wide variety of chelating Schiff bases can be made by changing the nature and position of the functional groups and introducing different sub-stituents into the reacting molecules. They provide a most useful class of compounds for the study of stereochemical aspects of metal chelates and of the reactivity of coordinated ligands.

Oxygen-containing multidentate ligands include the inorganic poly-phosphates and the organic hydroxycarboxylic acids. Complexes of these and their applications are considered in Chapter 8.

Novel types of chelating agents that have been recently developed are basically macrocyclic structures containing several oxygen, nitrogen or sulphur donor atoms. Each molecule is so constructed that a central cavity is formed by the rings containing the ligand atoms. This cavity is sur-rounded by an aromatic or aliphatic shell. The diameter of the cavity depends on the number of ligand atoms and the length of the carbon chains which link them. Provided the sizes of the cavity and a given metal ion are similar, the latter can be complexed and held quite firmly within the cavity.

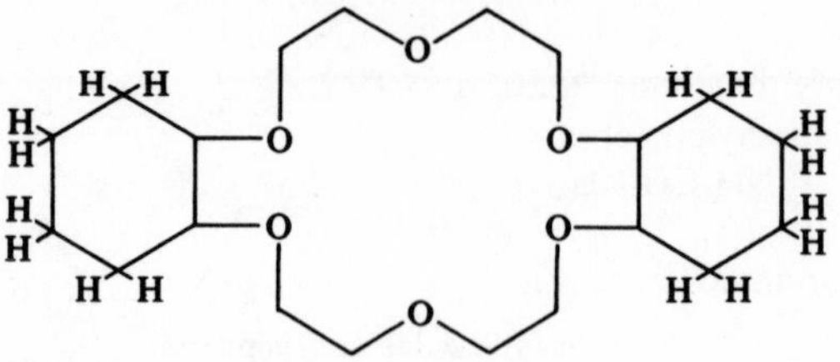

crown compound

As it is possible to synthesize molecules which have cavities of different sizes, this type of molecule is capable of high selectivity in its complexing action towards metals.

Macrocyclic ethers, commonly known as crown compounds, are exemplified by dicyclohexyl[18]crown-6. Here the macrocyclic ring is composed of 18 atoms, 6 of which are oxygens. Crown ethers are of special interest because they give rise to complexes of appreciable stability with ions of the alkali or the alkaline-earth metals.

When two tertiary nitrogens are introduced into a polyether structure, macrobicyclic diamines are formed. For example, in the compound whose structure is illustrated below, two nitrogen atoms are incorporated in the molecule and these are linked by three bridges, each composed of three (CH_2-CH_2) groups and two ether oxygens. The presence of two rings in the ligand creates a 'cage' structure within which a cation of appropriate size may be trapped. The cation is literally encapsulated within the cage of the molecular framework and for this reason such metal complexes are often known as 'cryptates' (derived from the Latin, *crypta*, meaning a hole or cavity).

In contrast to the polyethers, these diamines have basic properties and so their complexing ability is pH-dependent. However, their complexes with alkali metal ions are of unusually high stability compared with those of other ligands. They are of particular interest because selectivity towards ions like sodium or potassium can be developed by suitable variation in the size of the 'cage'.

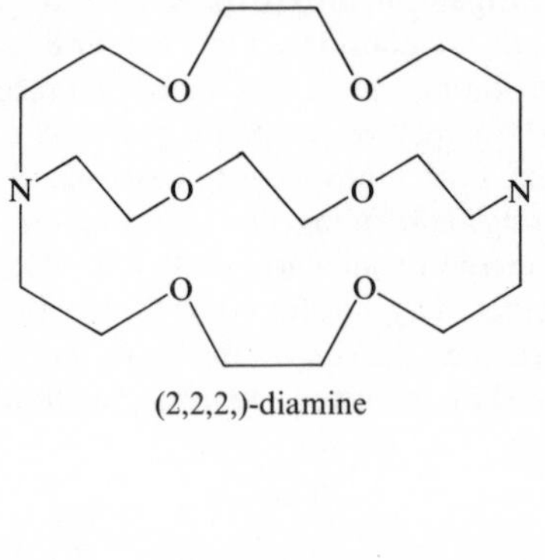

(2,2,2,)-diamine

PROBLEMS

1.1. Identify the donor atoms in each of the following ligands:

    (*i*)   triethylenetetraamine

               $NH_2CH_2CH_2NHCH_2CH_2NHCH_2CH_2NH_2$

    (*ii*)   uramildiacetic acid

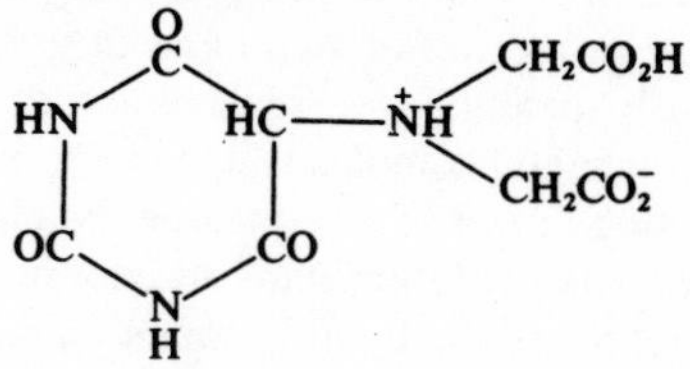

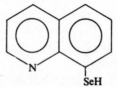

    (*iii*) 8-quinolineselenol

    (*iv*) diethylenetriaminepentaacetic acid

         $(HO_2C\ CH_2)_2NCH_2CH_2NCH_2CH_2N(CH_2CO_2H)_2$

                        |

                     CH_2CO_2H

    (*v*)   trifluoroacetylacetone $CF_3-CO-CH_2-CO-CH_3$

1.2. Comment on the relative magnitudes of the stability constants ($\lg K_1$) given below for complexes of iminodiacetic acid (IDA) and nitrilotriacetic acid (NTA).

Metal ion	Ni^{2+}	Cu^{2+}	Zn^{2+}	Cd^{2+}
$\lg K_{MIDA}$	8.21	10.55	7.03	5.35
$\lg K_{MNTA}$	11.53	12.96	10.67	9.83

1.3. What deductions may be made from the following data about the chelation of calcium ions by polyphosphates?

Ligand	$\lg K_1$
$P_3O_9^{3-}$ (trimetaphosphate)	0.62
$P_3O_{10}^{5-}$ (triphosphate)	4.6
$P_4O_{12}^{4-}$ (tetrametaphosphate)	3.77
$(P_nO_{3n+1})^{(n+2)}$ (polyphosphate)	7.35†

† for $n \simeq 60$ at pH = 7.

2. Properties of metals

As a complement to the preceding chapter, we shall now consider the typical properties of metal ions in so far as they affect chelate formation. Broadly speaking, we are dealing with the same properties that are significant for metal complexes in general, irrespective of whether the ligand is chelating or not. These include the electronic structure of the metal ion, its size and oxidation state, its coordination number and stereochemistry, the nature of the bond between metal and ligand (which may vary from essentially electrostatic to almost purely covalent), and, in the case of metals which exhibit variable valence, the relative stabilities of different oxidation states. For some metals, however, certain coordination numbers, stereochemistries, and oxidation states are exhibited only in their chelate complexes.

Classification of metal ions

Electronic structure

A convenient classification can be made in terms of the electronic structure of the ions. This is summarized in Table 2.1. The main differentiation to note here is between ions of types (i)-(iv), which have completed electronic subshells, and types (v) and (vi), which are characterized by incomplete subshells. The presence of an incomplete subshell confers special properties on the metal concerned, primarily that of variable valence. Moreover, the complexes of transition metals and the lanthanides and actinides are generally very much more stable than corresponding complexes of non-transition elements.

TABLE 2.1

Metals classified according to the electronic structures of their ions

Type	Outermost electrons	Examples
(i)	$1s^2$	Li^+, Be^{2+}
(ii)	ns^2np^6	Alkali metals, alkaline earths
(iii)	$(n-1)d^{10}$	Cu^+, Zn^{2+}, Ga^{3+}
(iv)	$(n-1)d^{10}ns^2$	Tl^+, Sn^{2+}, Pb^{2+}
(v)	$(n-1)d^1 \rightarrow (n-1)d^9$	Transition metals
(vi)	$(n-1)(f^1 \rightarrow f^{13})ns^2np^6$	Lanthanides ($n = 5$) and actinides ($n = 6$)

The properties of metal ions in relation to chelation are described later in this chapter on the basis of their electronic structures but, before doing so, two other classifications will be briefly considered.

Class (a) and (b) acceptors

An alternative division of metals, into two classes of acceptor ions (a) and (b), has been made by Ahrland, Chatt, and Davies (1958). The basis of this classification is the different affinities of the two classes towards ligands in aqueous solution.

Class (a) metals form their most stable complexes with the first ligand element of Groups V, VI, and VII, namely, nitrogen, oxygen, and fluorine. Most metals in their common valence states belong to this class. It includes, for example, the alkali and alkaline-earth metals, zinc, and the early metals of the transition series (up to Group VI).

The hydrogen ion is also a class (a) acceptor and therefore the stabilities of the chelates of class (a) acceptors tend to correlate approximately with the basicities of the ligands concerned.

In aqueous solution, the order of stabilities of a class (a) acceptor complexing with halide ions is $F \gg Cl > Br > I$. Hence this class shows a better coordinating power towards the more electronegative ligands.

Class (b) acceptors are those which form their most stable complexes with second row elements like phosphorus, sulphur, or chlorine or with elements below these in the periodic table. They are less numerous than class (a) acceptors. They include transition elements from Group VI onward, cadmium and mercury, and the heavy elements, thallium, lead, bismuth, and polonium. Class (b) acceptors in aqueous solution show the following order of affinities for halide ions: $F \ll Cl < Br < I$.

The ability of class (a) metals to form stronger complexes with fluoride than with the heavier halide or hydroxide ions indicates that the bonding within these is predominately electrostatic. The preference for this type of bond shows that the metal ions in question have weak polarizing power, this being a consequence of their electronic structures. Class (a) behaviour occurs with those ions in which the outermost shell contains 2 or 8 electrons (types (*i*) and (*ii*) above). Where the ion has a completely filled d-subshell (type (*iii*)), class (b) changes to class (a) character as the charge on the ion increases or, in cases (type (*iv*)) where there is an 'inert-pair' of s-electrons, class (b) behaviour predominates. For ions with incomplete d-subshells, class (a) character is shown when only one or two d-electrons are present but an increase in the number of d-electrons is accompanied by change to a class (b) metal. Clearly, class (b) behaviour is associated with the presence of a certain number of d-electrons; it is believed that these make an important contribution to the bonding in complexes of class (b) metals by interaction with empty π-orbitals on the ligands.

Class (b) metal ions are more strongly polarizing than those of class (a) and, as a result, covalent forces within their complexes contribute significantly to the metal-ligand bonding. The greater polarizing power of class (b) ions is consistent with their ability to form more stable complexes than class (a) metals with neutral molecules such as ethylenediamine. Correspondingly, the emphasis on electrostatic bonding in complexes of class (a) is illustrated by the stability of chelates formed from ligands containing two or more acidic functional groups, such as the di- and poly-carboxylic acids.

Hard and soft acids and bases

Metal ions behave as Lewis acids, and ligands as Lewis bases; the concepts of 'hardness' and 'softness' have been developed to describe systematically the interactions between them. For example, a 'hard' metal ion or acid is one which retains its valence electrons very strongly. It is not readily polarized and so is broadly equivalent to a class (a) metal. Hardness is a particular attribute of ions of small size and high charge, because this combination of properties results in low polarizability. Correspondingly, a 'soft' metal ion or acid is relatively large, does not retain its valence electrons firmly, and is easily polarized. 'Soft' metals include copper(I), silver(I), gold(I), thallium(I), palladium(II), and platinum(II); in fact, just those ions that show class (b) behaviour. Ligands containing highly electronegative donor atoms are difficult to polarize and can be classified as 'hard' bases. They include amines, ammonia, water, and ions like phosphate or sulphate. Easily polarized ligands, containing phosphorus, arsenic, or sulphur donor atoms, act as 'soft' bases.

As a general rule, the formation of stable complexes results from interaction between hard acids and hard bases, or soft acids and soft bases. The choice of chelating agent for complexing various metals can be rationalized on this basis. Thus a cyclic polyether contains a number of oxygen donors, which are not easily polarized, and it can therefore be classed as a 'hard' base forming stable complexes with 'hard' acids like K^+ and Ca^{2+}. Conversely, sodium diethyldithiocarbamate contains readily polarized sulphur atoms as donors and so is most suitable for the chelation of 'soft' acids like copper(I), cadmium(II), and mercury(II).

Ionic size, coordination number, oxidation state, and stereochemistry

The concept of ionic size is valuable for the rationalization and interpretation of the vast quantity of experimental data relating to metal complexes in which the bonding is primarily electrostatic. For an isolated ion, its radius is determined simply by the resultant force of attraction, after allowance has been made for the screening effect of intervening shells of electrons, between the nuclear charge and the outermost electrons. This quantity can be readily calculated. In condensed phases, we are always concerned with an ion in association with neutral molecules or ions of opposite charge and so there are additional attractive or repulsive forces operative that will affect its size. The size of an ion is not invariant but depends on its coordination number, the nature of the molecules or ions with which it is linked, and the extent to which the bonding departs from the purely electrostatic because of polarization effects. In the case of transition-metal ions, certain numbers of d-electrons can give rise to either high-spin (HS) or low-spin (LS) configurations according to the magnitude of the surrounding ligand field. This is an additional factor which affects the size of these ions.

TABLE 2.2
Effective radii r (pm) of metal ions in octahedral coordination†

	Group I		Group II		Group III		Group IV
	Li^+	74	Be^{2+}	35	B^{3+}	23	
	Na^+	102	Mg^{2+}	72	Al^{3+}	53	Si^{4+} 26
	K^+	138	Ca^{2+}	100	Sc^{3+}	73	Ti^{4+} 60
	Rb^+	149	Sr^{2+}	116	Y^{3+}	89	Zr^{4+} 72
	Cs^+	188	Ba^{2+}	136	La^{3+}	106	Hf^{4+} 71
	Cu^+	96	Zn^{2+}	74	Ga^{3+}	62	Ge^{4+} 54
	Ag^+	115	Cd^{2+}	95	In^{3+}	79	Sn^{4+} 69
	Au^+	137	Hg^{2+}	102	Tl^{3+}	88	Pb^{4+} 77

Transition metals

	Group I	Group II	Group III	Group IV
	Ti^{2+} 86	V^{2+} 79	Ni^{2+} 70	
LS	73	67	61	65
	Cr^{2+}	Mn^{2+}	Fe^{2+}	Co^{2+}
HS	82	82	77	73
	Ti^{3+} 67	V^{3+} 64	Cr^{3+} 62	
LS	58	55	52	56
	Mn^{3+}	Fe^{3+}	Co^{3+}	Ni^{3+}
HS	65	64	61	60

Lanthanides

La^{3+} 106.1	Ce^{3+} 103.4	Pr^{3+} 101.3	Nd^{3+} 99.5
Pm^{3+} 97.9	Sm^{3+} 96.4	Eu^{3+} 95.0	Gd^{3+} 93.8
Tb^{3+} 92.3	Dy^{3+} 90.8	Ho^{3+} 89.4	Er^{3+} 88.1
Tm^{3+} 86.9	Yb^{3+} 85.8	Lu^{3+} 84.8	

Actinides

Ac^{3+} 118	Pa^{3+} 113	U^{3+} 106	Np^{3+} 104
Pu^{3+} 100	Am^{3+} 101	Cm^{3+} 98	

† From Shannon and Prewitt (1969). *Acta crystallogr.* B **25**, 925

Sets of ionic radii have been calculated *inter alia* by Pauling, Goldschmidt, Zachariasen, and Ahrens. Differences in the values assigned are found because of the differing assumptions underlying the calculations. The most useful figures are those which reproduce most nearly the inter-ionic distances observed in crystalline compounds. On this criterion, one of the best sets of 'effective' ionic radii is that derived by Shannon and Prewitt, on the basis of about 100 measured interionic distances in oxides and fluorides. Their values, given in Table 2.2, refer to a coordination number of 6.

By simple geometry, the larger the metal ion, the greater the number of ligand atoms which can surround it so that each may still be regarded as in contact with the central ion. Low coordination numbers are therefore to be expected for ions like Li^+, Be^{2+}, and ions of low oxidation state in the first transition series. Correspondingly, high coordination numbers are found for the heavier alkali and alkaline-earth metal ions, the lanthanides and the actinides. For monatomic anionic ligands, the limiting size of the metal ion for a specified coordination number can be estimated by calculation of the radius ratio (radius of metal ion/radius of ligand). This purely geometrical approach cannot be directly applied to metal chelates unless, which is not generally the case, the 'size' of the donor atoms is accurately known.

The charge on the metal ion also influences the coordination number. If this charge is low, say +1 or +2, then the acceptance of only a few lone pairs of electrons from a small number of ligands could lead to the accumulation of enough negative charge on the metal to prevent the coordination of additional ligands. When the charge is high, for example +3 or +4, then a high coordination number can be reached without too much negative charge building up on the metal. Metals in very high oxidation states (+6 and +7) generally show low coordination numbers because the small ionic size makes it sterically impossible for a large number of ligand atoms to be coordinated. In the case of complexes where the bonding between metal and ligand is primarily covalent, the coordination number of the metal is determined by the number of σ-bonding orbitals available on the metal for combination with ligand orbitals of the same symmetries and similar energies. The most complete description of the bonding within such complexes is given by molecular orbital theory. All possible interactions between metal and ligand orbitals are considered and calculations permit quantitative estimates of the electronic energies within the molecular orbitals formed.

Wherever we are concerned with a metal atom or ion which has a spherically symmetrical core of non-bonding electrons, as in types (*i*), (*ii*), and (*iii*) ions in Table 2.1, the arrangement of ligands is that of maximum symmetry, irrespective of whether the bonding is electrostatic or covalent. For example, a 4-coordinate complex is tetrahedral, a 5-coordinate complex is trigonal bipyramidal, a 6-coordinate complex octahedral, and so on. Where the non-bonding core is non-spherical, for example in transition-metal ions, distortions from maximum symmetry occur for asymmetric distributions of electrons (Jahn-Teller effect). Thus tetragonal distortion within complexes of a d^9 HS ion like Cu^{2+}, and of a d^8 LS ion like Ni^{2+}, effectively results in square-planar stereochemistry around the metal. In special cases, the shape of the complex is determined primarily by the stereochemistry and electronic characteristics of the ligand, for example, some bidentate sulphur-containing ligands force trigonal prismatic stereochemistry on a 6-coordinated metal ion instead of the more symmetrical octahedral arrangement.

Stereochemistry of ions with different types of electronic structure

Type (i) ions

Lithium and beryllium head the first two groups of the periodic table. Both elements have unusual properties which distinguish them from other metals in their respective subgroups. These are primarily the result of the high ratio of charge to ionic size which means the ions have appreciable polarizing power and hence that many of their compounds have significant covalent contributions to the bonding.

The ions Li^+ and Be^{2+} are not 'hard' acids and the chelating ligands reacting with them characteristically contain oxygen and/or nitrogen donors. Typically for first-row elements, their coordination number is limited to a maximum of 4.

Lithium is chelated in its adduct LiOx.HOx (with 8-hydroxyquinoline HOx), and in other complexes LiL, formed from ligands (HL), such as o-nitrophenol and 1-nitroso-2-naphthol. Other alkali-metal ions form similar complexes.

Beryllium chelates include bis(acetylacetonato)beryllium, containing tetrahedrally coordinated metal and two planar chelate rings.

bis(acetylacetonato)beryllium

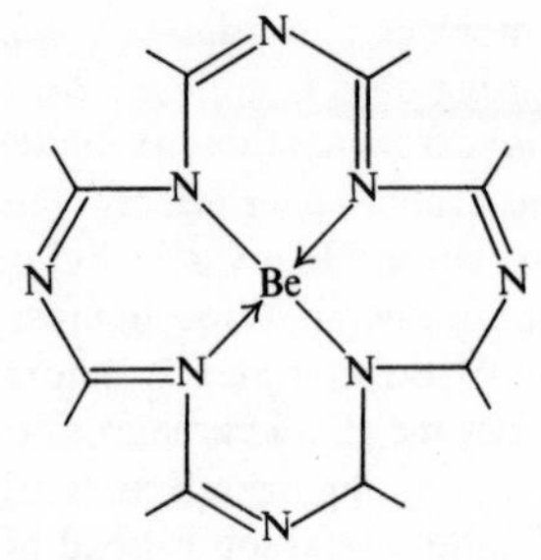

beryllium–phthalocyanine complex

Although it is frequently found in coordination with oxygen donors, there are only a few instances known where beryllium forms a complex containing Be–N bonds. One example is the beryllium-phthalocyanine complex. This is of special interest because the planar geometry typical of phthalocyanines forces the beryllium atom to adopt a planar configuration instead of its normal tetrahedral stereochemistry.

Type (ii) ions

This group includes the alkali metals, the alkaline-earth metals, the tripositive ions of aluminium, scandium, and yttrium, and tetrapositive silicon.

The alkali metals have low first-ionization energies and easily lose an electron to form large, univalent cations of noble-gas electronic configuration. The shielding effect of the closed shell (ns^2np^6) is large and the ions therefore have only small polarizing power. These elements consequently show predominantly electrostatic bonding in their compounds. They are typical hard acids. The best-known chelates are with oxygen-containing ligands.

For many years, the known chelates of the alkali metals were limited to those with ligands like salicylaldehyde or benzoylacetone. Coordination of one anion is sufficient to produce an uncharged complex, but coordination of further neutral molecules of the ligand or of solvent molecules permits expansion of the coordination number to 4 or 6. Thus rubidium coordinates two salicylaldehyde molecules as well as one anion to attain a coordination number of 6.

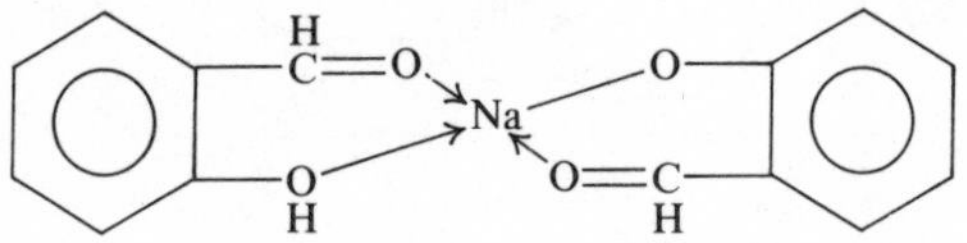

sodium chelated by salicylaldehyde

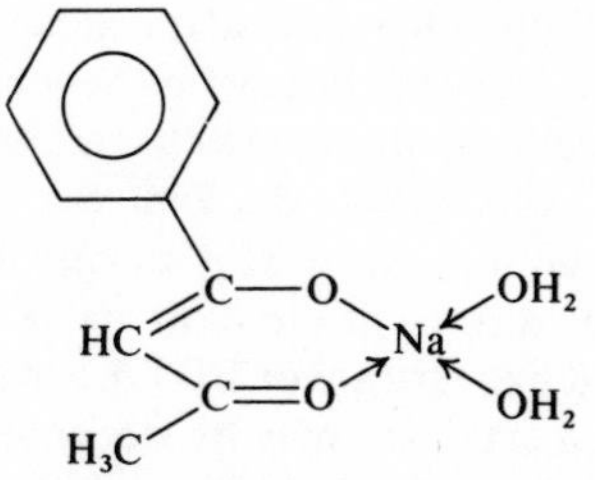

sodium chelated by benzoylacetone

In 1967, the coordination chemistry of the alkali metals was revolutionized by the synthesis of their chelates with macrocyclic polyethers. These complexes are of special interest, because of their structural similarities to sodium and potassium complexes with some naturally occurring antibiotics, and in relation to the vital roles played by sodium and potassium ions in biological systems. Coordination numbers of 8 or 10 have been found in some of the complexes with polyethers.

The alkaline-earth metals form divalent cations of noble-gas electronic configuration. They are similar to the alkali metals in acting as hard acids and in preferentially forming complexes with oxygen donors. In many instances, for example, in their oxalate and glycinate complexes (Table 2.3), the order of stability of the chelates varies in the inverse order of the cationic radii. This is suggestive of an interaction between metal and ligand which is primarily electrostatic, for the smaller the size of an ion, the greater its attraction for a negatively charged ligand and hence the more stable the resulting complex. Although the stability sequence $Ca^{2+} > Sr^{2+} > Ba^{2+}$ is preserved for tartrate, the magnesium complex shows an unexpectedly low stability. Similarly, for complexes of EDTA (H_4Y), lg $K(MgY)$ (= 10.3) is less than lg $K(CaY)$ (= 12.1). For these complexes the data relate to aqueous solution and so are a quantitative

TABLE 2.3
Stability constants (lg K_1)
for complexes of carboxylic acids
and alkaline-earth metal ions

Ligand	Metal			
	Mg^{2+}	Ca^{2+}	Sr^{2+}	Ba^{2+}
Oxalate	3.43	3.0	2.54	2.31
Glycinate	3.44	1.43	0.91	0.77
Tartrate	1.36	1.80	1.65	1.62

expression of the stability of the particular complex relative to that of the hydrated metal ion. Because the interaction between metal ion and water is primarily a electrostatic ion-dipole effect, this effect is greater — for ions of constant charge — the smaller the ionic size. Of the four metals in question, we would therefore expect Mg^{2+} to form the most stable hydrate. It may not necessarily also form the most stable complex with a multi-dentate anionic ligand like tartrate or EDTA because coordination of all donor atoms to such a small ion may be sterically impossible. Alternatively, if all donors are coordinated, then this may involve more strain (reflected in a lower stability) in the ligand structure, the smaller the cation involved.

Like beryllium, magnesium forms complexes with nitrogen donors already linked in a ring system. The best-known example of these is the coordination of magnesium in the chlorophyll pigments by four coplanar pyrrole nitrogens.

Aluminium, scandium, and yttrium in Group III and silicon in Group IV show generally class (a) behaviour in their preferential coordination with oxygen-containing ligands. As for the two preceding Groups, chelation of Group III metals involving nitrogen takes place with ligands such as 8-hydroxyquinoline, imidodiacetic acid, and EDTA. The smaller Sc^{3+} ion forms more stable complexes than Y^{3+} (cf. lg $K(ScY) = 23.1$ and lg $K(YY) = 18.09$), indicative again of the importance of electrostatic bonding. The aluminium complexes of multidentate ligands like EDTA show a lower stability (lg $K(AlY) = 16.13$) than those of scandium. This is in keeping with the small size of the tripositive aluminium ion which does not permit coordination of all possible donor atoms on the ligand.

Type (iii) ions

These ions have a d^{10}-subshell of electrons. This electronic structure is characterized by the relatively high polarizing power of the ion and so the bonds formed by type (iii) elements tend to have a significant covalent character. A measure of the polarizing power of an ion is its electron affinity — the energy released when one or more electrons are taken up to give the atom. This is, of course, a quantity of the same magnitude as the ionization energy of the atom but of opposite sign. The electron affinities of type (iii) ions are much greater than those of comparable ions of type (ii). For example, the electron affinity of K^+ is 418.4 kJ mol^{-1} whilst that of Cu^+ is 745.2 kJ mol^{-1}; that of Ca^{2+} is 1734.7 kJ mol^{-1} compared with 2639.3 kJ mol^{-1} for Zn^{2+}.

The increased polarizing power of the ion is accompanied by a decrease in the coordination number commonly observed. Thus the usual coordination number found for copper(I) is 4; as is usually the case for non-transition-metal ions, this coordination number is associated with tetrahedral

4-coordinate dithiolene chelate

stereochemistry. Ligands containing oxygen donors do not form very stable complexes with copper(I) and it preferentially chelates with sulphur-containing ligands. In the oxidation state of $+1$, copper clearly shows class (b) properties. Copper(II) has some characteristics of both classes.

The preferred coordination number for silver(I) is only 2. This is associated with a linear arrangement of bonds. Chelation is rarely observed with Ag^+ ions simply because most chelating agents are structurally unable to coordinate to give such a disposition of bonds. Those chelate complexes which are known usually contain at least one sulphur donor atom, indicating the class (b) character of silver in this oxidation state.

Similarly, gold(I) has class (b) properties. It shows a coordination number of 4 and tetrahedral stereochemistry in complex ions such as $[Au(diars)_2]^+$ (diars = o-phenylenebis(dimethylarsine)).

The metals of Group IIB, zinc, cadmium, and mercury, show considerably more covalent character in their complexes than the alkaline earths, and as a consequence, the relatively low coordination number of 4 is commonly found.

Zinc(II) shows predominantly the attributes of a class (a) metal in forming stable complexes with oxygen and nitrogen donors. Some indications of class (b) properties are also found, particularly in its complexes with sulphur-containing ligands. An example of the latter is the 4-coordinate dithiolene chelate.

The less common coordination number of 5 is found in a number of chelates of Group IIB metals. This occurs, for example, with a tridentate ligand like 2,2′,6′,2″-terpyridyl (terpy) or pyridine-2-aldehyde 2′-pyridyl-hydrazone (paphy). This occupies three coordination positions around the metal and two more can be filled by, for example, unidentate ligands like chloride, so that a neutral 5-coordinate species is obtained. Thus the complexes $Zn(terpy)Cl_2$ and $Zn(paphy)Cl_2$ both contain zinc in a distorted trigonal bipyramidal stereochemistry. In such cases, the stereochemistry and coordination number adopted are a direct consequence of the configuration of the tridentate ligand.

Cadmium(II) is borderline between class (a) and (b) metals whilst mercury(II) is in class (b). As with zinc, the coordination numbers are generally 4 and 6. Mercury is tetrahedrally coordinated by arsenic atoms in

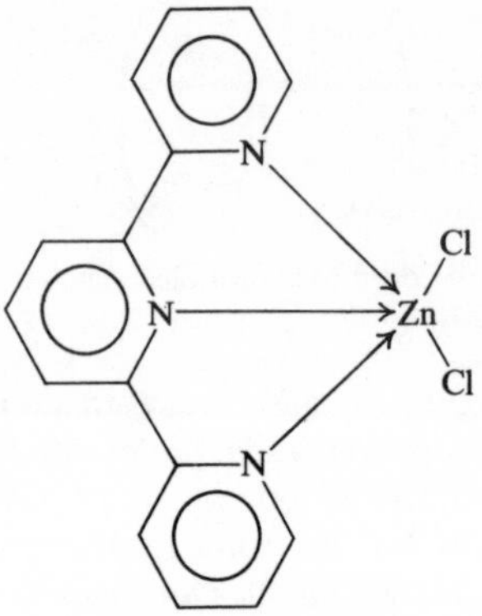

dichloro (terpy) zinc

$[Hg(diars)_2]^{2+}$ and by sulphur and nitrogen in its dithizone complex, $HgDz_2$. Coordination numbers of 5, 6, and 8 have been variously reported in other complexes.

In Group IIIB, the lighter elements, gallium and indium, show predominantly class (a) behaviour in the tendency of the tripositive ions to interact with oxygen-containing chelating agents like oxalic acid or mixed oxygen-nitrogen donors such as 8-hydroxyquinoline. As in the preceding Group, class (b) behaviour is most evident with the heaviest element, thallium, in its +3 oxidation state. Stable complexes are, however, known with oxygen and nitrogen donors and so it is best to regard thallium(III) as borderline between the two classes.

Type (iv) ions

These ions contain a pair of s-electrons outside a completed electronic shell. This pair does not participate in bonding in many compounds of the element in question and is often described as 'inert'. As a result, there exists a well-defined oxidation state which is two units lower than the Group valence of the element. Thus a range of thallium(I) compounds exists: similarly tin(II) and lead(II) compounds are well known.

Tl^+ and K^+ ions have very similar ionic sizes and superficially there are some resemblances between compounds containing these. However, Tl^+ contains a filled 5d-subshell and shows a much greater polarizing power than K^+. In its compounds, it therefore exhibits a greater degree of covalent bonding.

Thallium(I) shows class (a) behaviour in, for example, its chelates with 8-hydroxyquinoline and β-diketones. Class (b) characteristics are evident also. Thus thallium is quite unlike the alkali metals in its ability to complex with sulphur in ligands like dithizone.

Tin(II) and lead(II) show class (b) behaviour in their preferential chelation with sulphur-containing ligands. Again covalent bonding is important in their chelates and the lower coordination numbers of 4, 5, or 6, are commonly found. Coordination with sulphur is illustrated in the case of tin(II) by its bis-complex with toluene-3,4-dithiol and of lead(II) with its bis-complex with dithizone. The formation of these particular chelates is the basis of widely used analytical procedures for these two metals.

Type (v) ions

In the three series of transition metals, each d-subshell is progressively filled with electrons as the atomic number increases. The metals early in each series have few electrons outside the noble-gas electron configuration, ns^2np^6. The shielding effect of this configuration is good, as we have seen in the case of the alkali and alkaline-earth metals, and so the ionization energies are relatively low. It is not surprising therefore that ions with a d^0-configuration, such as Ti^{4+} and Nb^{5+}, show similarities in their coordination behaviour to the metals of Groups IA and IIA. This is broadly true as regards preferential coordination with oxygen donors. The tendency towards covalent character in the bonding is more pronounced because such highly charged ions possess considerable polarizing power.

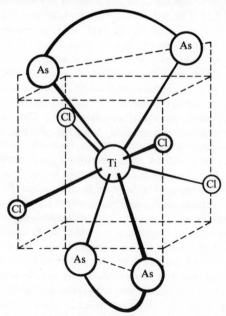

FIG. 2.1. Dodecahedral stereochemistry of Ti(diars)₂Cl₄.

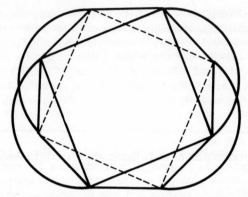

FIG. 2.2. Square anti-prismatic stereochemistry of $Zr(acac)_4$.

The early transition metals have relatively large ions and their sizes, together with the high charges, account for the prevalence of high coordination numbers found in their chelates. Many instances of 7- and 8-coordination are known for the metals of Groups IVA, VA, and VIA.

One of the first examples of 8-coordination for elements of the first transition series was found in the titanium(IV) complex with o-phenylene-bis(dimethylarsine), $Ti(diars)_2Cl_4$. The complex has dodecahedral stereochemistry (Fig. 2.1). This is the favoured stereochemistry for 8-coordinate complexes where four ligands of one kind and four of a second kind are present. Thus the zirconium, hafnium, and vanadium(IV) analogues of the titanium 'diarsine' complex have the same structure.

For the case of four small bidentate ligands, the dodecahedral structure is again found, as in tetraoxalatozirconate(IV), $Zr(C_2O_4)_4^{4-}$: The alternative structure of a square antiprism is found in complexes like $Zr(acac)_4$, where acac represents the acetylacetonate anion. This stereochemistry appears to be favoured with relatively large ligands (Fig. 2.2).

Across a transition-metal series, d-electrons are successively introduced into the penultimate electronic shell. These have poor shielding power, so the effective nuclear charge of the ions increases and so does their electron affinity. There is a concomitant decrease in ionic radius. The net effect of these changes is an increase in polarizing power of the ion as the number of d-electrons increases, an increase in covalent character of the bonding, and a decrease in maximum coordination number to 6. Consequently, the character of the transition metal changes gradually from class (a) for the early members of each series to class (b) for those at the end.

The complexes formed by many chelating ligands, in which the donors are either oxygen or nitrogen atoms, with stable divalent ions of the first transition series show the following order in their stability constants: Mn < Fe < Co < Ni < Cu > Zn. This, known as the Irving-Williams order, is illustrated by complexes of ethylenediamine (Chapter 1) and EDTA (Chapter 5). As the atomic number increases changes in certain properties of the metal ion, such as its size and electron affinity, provide the key to understanding this particular sequence. Thus, the radius of the high-spin divalent ions in octahedral coordination follows the same sequence for all ions except one. The situation for the exception, Cu^{2+}, is not clear-cut. Its radius cannot be unequivocally assigned because it does not, due to the Jahn-Teller effect, occur in a regular octahedral environment. However, increase in stability is roughly paralleled by decrease in ionic size, a trend to be expected·in view of the contribution to bonding of the electrostatic interaction between a cation and a ligand which is either anionic or a polar molecule. An approximately linear correlation exists also between the electron affinities of the various ions as measured by the second ionization energy of the metal, and the stability constants of their complexes with a particular ligand. This correlation is entirely in keeping with the simple picture of coordinate bonding being due to the donation of electrons from the ligands to the metal ion, the strength of the metal-ligand bond naturally increasing with the electron-attracting power of the cation. Of course, in most chelates, it is incorrect to think in terms solely of either electrostatic or covalent bonding; from the nature of the components both types of interaction are normally contributory.

Some contribution to the stability of the complexes of certain transition metals comes from crystal-field stabilization energy (CFSE).[†] The formation of a metal chelate in aqueous solution involves the replacement of water molecules by ligands which are generally higher in the spectrochemical series (that is, they produce greater crystal-field splitting than water). In the sequence of octahedrally coordinated ions $Mn^{2+} \rightarrow Zn^{2+}$, the CFSE is zero for Mn^{2+} (d^5) and Zn^{2+} (d^{10}) but for intermediate ions rises to a maximum with Ni^{2+} (d^8). Although the CFSE represents only a small fraction of the total energy of formation of a complex, nonetheless for those ions which are so stabilized it can increase the stability constant of their complexes by about 5-10 per cent above the value predicted if crystal-field effects were absent.

The Irving-Williams order is not invariably followed and a number of factors operate to bring about departures from it. The case of phenanthroline complexes is particularly interesting. The sequence of stabilities of the 1:1 complexes between divalent ions and this ligand is the Irving-Williams order. On the other hand, the overall stability constants for the tris complexes of the same ions are:

	Mn^{2+}	Fe^{2+}	Co^{2+}	Ni^{2+}	Cu^{2+}	Zn^{2+}
lg $K_1K_2K_3$	10.3	21.3	19.9	24.8	20.8	17.55

[†]See also Earnshaw and Harrington: *The chemistry of the transition elements* (OCS 13).

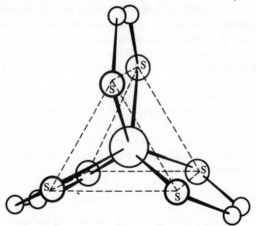

1,2-dithiolenes

First, we note the extra stability of $[Fe(phen)_3]^{2+}$. This has been related to the changeover from high-spin to low-spin Fe^{2+} during the course of chelation with three phenanthroline molecules. This is caused by the strong ligand-field effect exerted by the phenanthroline ligand. Fe^{2+} is a d^6 ion and its low-spin state is actually diamagnetic with all six electrons paired up in the lower-energy t_{2g} orbitals. In this state, the effective radius (Table 2.2) is significantly smaller than that of high-spin Fe^{2+}. Increased stability of the 1:3 complex results from all electrons now being located in the lower-energy orbitals (due allowance being made for the energy required to force the unpaired electrons to pair up in these orbitals) and the smaller size of the low-spin Fe^{2+} ion.

Secondly, although according to the Irving-Williams order Cu^{2+} should form the most stable complexes of all the divalent ions in the first series, its tris(phenanthroline) complex is noticeably less stable than that of nickel. There is a considerable drop in stability as the phenanthroline molecules are successively complexed by copper (cf. lg $K_1 = 9.0$; lg $K_2 = 6.7$; lg $K_3 = 5.1$). Although stepwise constants usually decrease with increase in the number of ligands, both $[Cu(phen)_2]^{2+}$ and $[Cu(phen)_3]^{2+}$ show anomalously low stabilities. Now Cu^{2+} cannot exist in a regular octahedral environment because of Jahn-Teller distortion and in many complexes it is

FIG. 2.3. Trigonal prismatic structure of Re(std)₃.

effectively 4-coordinate with a square-planar distribution of bonds. However, the $[Cu(phen)_2]^{2+}$ ion must have a more irregular structure than this because square-planar stereochemistry would be unstable because of steric hindrance between hydrogen atoms in positions 2 and 9 in the two coordinated ligands. In the case of $[Cu(phen)_3]^{2+}$, reduced stability is the result of two conflicting factors; on the one hand, the operation of a Jahn-Teller effect tending to give a distorted stereochemistry and, on the other, the symmetry and rigidity of the ligand which should favour a more regular structure with six more or less equivalent Cu$-$N bonds.

For many years after the establishment by Werner of the octahedral stereochemistry of a 6-coordinate transition-metal ion, it was believed that no alternative stereochemistry could exist. However, since 1965, several 6-coordinate complexes have been synthesized which have a trigonal prismatic arrangement of ligand atoms around the metal. The ligands which give rise to this unusual structure are the 1,2-dithiolenes. The first complex shown by X-ray diffraction to have a trigonal prismatic structure was $Re(sdt)_3$ (sdt = $C_6H_5C(S) = C(S)C_6H_5^{2-}$). The rhenium atom is surrounded by six sulphur atoms in the almost exactly trigonal prismatic arrangement shown in Fig. 2.3. The sides of the prism are square and the five-membered ReS_2C_2 chelate rings are planar and radiate outwards from the three-fold axis of the trigonal prism in a regular 'paddle-wheel' manner. This geometry appears to be quite stable because the spectroscopic properties of the complex are substantially unchanged on solution. The distance between non-bonded sulphur atoms is relatively short (305 pm). This has been interpreted in terms of the presence of interligand forces which are considerably stronger than those in octahedral complexes. The strength of these interactions between coordinated ligand molecules is believed to be a major factor in the stabilization of the unusual stereochemistry.

Square-planar geometry is commonly observed for complexes of 4-coordinated metal ions of d^8 (Au^{III}, Ni^{II}, Pd^{II}, and Pt^{II}) and d^9 (Cu^{II} and Ag^{II}) configurations. This stereochemistry is found, for example, in copper(II) dithizonate, bis(pyridine-2-carboxylato)silver(II), and bis-(dimethylglyoximato)nickel(II).

bis(pyridine-2-carboxylato)silver(II)   bis(dimethylglyximato)nickel(II)

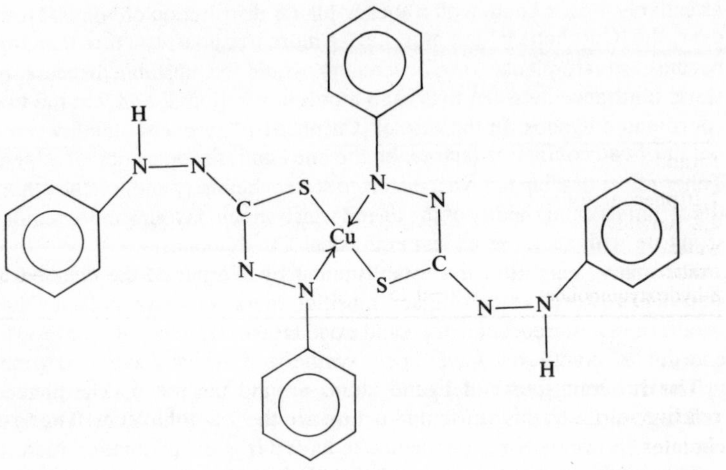

copper(II)dithizonate

Redox properties†

Complexation of a transition-metal ion results in significant changes in the electrode potential existing between two different oxidation states of the element.

The ion half-reaction linking two oxidation states is given generally by:

$$\text{oxidized chelate} + z\text{e} = \text{reduced chelate}$$
$$\text{(Ox)} \qquad\qquad \text{(Red)}$$

and the electrode potential by:

$$E = E^0 + \frac{RT}{zF} \ln \frac{a(\text{Ox})}{a(\text{Red})}$$

where $a(\text{Ox})$ and $a(\text{Red})$ are the relative activities of the two forms and E^0 is the standard electrode potential, relating to unit relative activity, of each form at 298 K. z is the number of electrons transferred in the redox reaction. E^0 is related to the standard free energy change ΔG^0 for the ion half-reaction:

$$\Delta G^0 = -zFE^0.$$

†The background to this section will be found in Robbins's *Ions in solution* (2): *electrochemistry* (OCS 2); see also Pass's *Ions in solution* (3): *inorganic properties* (OCS 7.)

TABLE 2.4

*Standard electrode potentials ($E°/V$) for chelates
of the Fe(II) and Fe(III) redox couple*

Ligand	$E°/V$	Ligand	$E°/V$
1,10 phenanthroline	+1.20	cytochrome f	0.4
2,2′ bipyridyl	+1.096	cytochrome c	0.25
water	+0.77	iron protoporphyrin	−0.12
oxalate	−0.01	horseradish peroxidase	−0.3
8-hydroxyquinoline	−0.15		

The different magnitudes of the E^0 values in Table 2.4 reflect different relative stabilities of the +2 and +3 oxidation states of iron in its various chelates.

The relationship between electrode potential and stability constant may be clarified by an example:

For $Fe(aq)^{3+} + e = Fe(aq)^{2+}$ $\qquad \Delta G^0 = -FE_1^0$

$\qquad Fe(aq)^{2+} + 3\,phen = Fe(phen)_3^{2+}$ $\qquad \Delta G^0 = -2.303\,RT \lg \beta_3''$

$\qquad Fe(phen)_3^{3+} = Fe(aq)^{3+} + 3\,phen$ $\qquad \Delta G^0 = -2.303\,RT \lg \beta_3'''$,

where

$$\beta_3'' = \frac{\{Fe(phen)_3^{2+}\}}{\{Fe(aq)^{2+}\}\{phen\}^3} = 10^{21.3}$$

$$\beta_3''' = \frac{\{Fe(phen)_3^{3+}\}}{\{Fe(aq)^{3+}\}\{phen\}^3} = 10^{41.1}.$$ Brackets indicate activities.

By addition, $Fe(phen)_3^{3+} + e \rightleftharpoons Fe(phen)_3^{2+}$.
For this, $\Delta G'^0 = -FE_2^0 = -FE_1^0 + 2.303\,RT\,(\lg \beta_3''' - \lg \beta_3'')$ and $\therefore$

$$E_2^0 = E_1^0 + \frac{2.303\,RT}{F}\,(\lg \beta_3'' - \lg \beta_3''').$$

Since

$$E_1^0 = +0.77$$

$$E_2^0 = 0.77 + \frac{2.303 \times 8.314\ \text{JK}^{-1}\ \text{mol}^{-1} \times 298\ \text{K}\ (21.3 - 14.1)}{9.65 \times 10^4\ \text{C mol}^{-1}}$$

$$= 1.20\ \text{V}.$$

Here, and in the case of bipyridyl ligand, the more positive value of E^0 (compared with the aquo system) is a consequence of the greater stability of $[Fe(phen)_3]^{2+}$ compared with $[Fe(phen)_3]^{3+}$. In other words, chelation by phenanthroline results in greater stability of the lower oxidation state. This is due to the stabilization energy released when the low-spin $[Fe(phen)_3]^{2+}$ complex is formed. In contrast, $[Fe(phen)_3]^{3+}$ contains a high-spin d^5-ion which has no ligand-field stabilization energy.

In the case of oxalate, 8-hydroxyquinoline, and other similar ligands, coordination involves a negative ligand ion with a positively charged metal ion. Relative to the formation of an aquo complex, there is a favourable enthalpy of formation of the chelates of both Fe(II) and Fe(III). Simply in terms of the electrostatic interaction between the charged species, this enthalpy change will be more favourable for the Fe^{3+} ion because it carries the greater charge. Hence the enhanced stability of the iron(III) chelate.

Type (vi) ions

Cations of the lanthanide elements show class (a) behaviour. The best-known complexes are those with chelating agents containing oxygen donors. Coordination with nitrogen donors is usually found only in association with oxygen donors (for example, in oxine and EDTA complexes). The reason for this is that the lanthanide ion forms more stable bonds with oxygen and so, in the presence of ligands which donate solely via nitrogen, there is a strong tendency in aqueous solution for preferential bond-formation with solvent oxygen atoms. However, in non-aqueous media, for example in acetonitrile solution, complexes of considerable stability with nitrogen donors have been made.

The relatively large size of the lanthanide ions allows their coordination number to be as high as 8 or 9, or, in a few cases, 10. For example, lanthanum is 8-coordinate in $La(phen)_4(ClO_4)_2$; europium is 8-coordinate in $Eu(acac)_3(phen)$; cerium is 8-coordinate in $Ce(acac)_4$; and lanthanum is 9-coordinate in $La(terpy)_3(ClO_4)_3$, and 10-coordinate in $La(H_2O)_4$ (H-EDTA), where H-EDTA represents monoprotonated sexadentate EDTA.

The interaction between lanthanide ion and ligand is largely electrostatic. In accordance with this, the stability of complexes with a common ligand increases with decreasing size of the M^{3+} ion. In the lanthanide series, the filling-up of the 4f-subshell is accompanied by a gradual decrease in radius of the tripositive ion (the lanthanide contraction). The largest ion is La^{3+} (radius = 106.1 pm) and the smallest is Lu^{3+} (radius = 84.8 pm). Consequently, the stability of lutetium complexes is greatest and that of lanthanum complexes is least within the lanthanide series. For example, in the case of EDTA complexes, lg $K_{LaY} = 15.5$ whereas lg $K_{LuY} = 19.83$. Differences in stability of lanthanide complexes have been exploited very effectively in separation procedures for the isolation of individual lanthanides in a state of high purity.

Actinide elements generally resemble the lanthanides in their complexes. High coordination numbers are again prevalent. For example, 8-coordination is found in the acetylacetonates of thorium(IV) and uranium(IV), Th(acac)$_4$ and U(acac)$_4$. In both molecules, the eight oxygens are arranged around the metal in an almost regular square antiprism. In their high oxidation states, several actinides form stable oxocations and the stereochemistry of their complexes shows special features arising directly from the presence of these ions. For example, uranyl nitrate hexahydrate $UO_2(NO_3)_2.6H_2O$ contains 8-coordinate uranium, but neither of the more commonly found dodecahedral or square antiprismatic configurations is observed. Instead, the linear uranyl ion UO_2^{2+} is surrounded equatorially by an irregular hexagon of six oxygen atoms, four from two bidentate nitrato groups and two from water molecules. In this diagram the dotted lines represent bonds in UO_2^{2+} which are perpendicular to the plane of the paper.

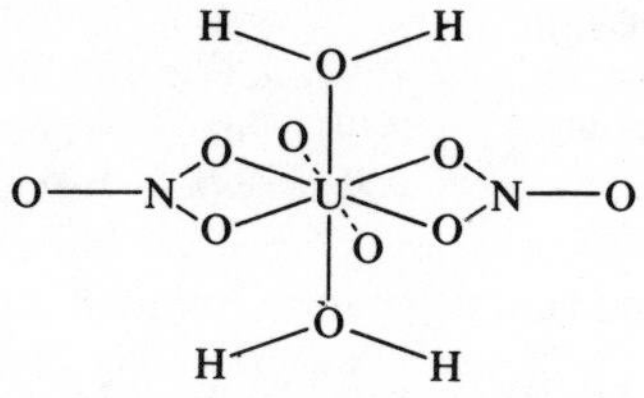

uranyl nitrate hexahydrate

PROBLEMS

2.1. In a number of enzymes, catalytic activity is associated with the $-SH$ groups of cysteine residues. Which of the following metal ions would you expect to inhibit the catalytic activity of such enzymes?
(*i*) Na^+ (*ii*) Pb^{2+} (*iii*) Hg^{2+} (*iv*) Ca^{2+} (*v*) Cd^{2+}

2.2. What stereochemistry would you expect to be shown by copper(II) in the following complexes?
[Cu N(CH$_2$CH$_2$NH$_2$)$_3$NCS]SCN, bis(acetylacetonato)copper(II), and [Cu(bipy)$_2$I]I (bipy = 2,2′-bipyridyl).

2.3. Comment on the trends in the following stability-constant data:

Metal-Ligand	lg K_1	lg K_2	lg K_3
Cu-en	10	9	−0.9
Cu-bipy	8.0	5.6	3.5

Why is lg K_3 so small with ethylenediamine as ligand compared with 2,2'-bipyridyl?

2.4. For the EDTA complex of Fe^{3+}, $Fe(EDTA)^-$, lg $K_1 = 25.1$, and for the redox reaction:

$$Fe(EDTA)^- + e = Fe(EDTA)^{2-}, \quad E^0 = +0.133 \text{ V}.$$

Calculate the stability constant of the Fe^{2+} complex $Fe(EDTA)^{2-}$, using $E^0 = 0.77$ V for the $Fe(aq)^{3+}/Fe(aq)^{2+}$ couple, $R = 8.314$ J K mol^{-1}, $T = 298$ K and $F = 9.65 \times 10^4$ C mol^{-1}.

2.5. In the redox half-reaction:

$$Fe(phen)_3^{3+} + e = Fe(phen)_3^{2+},$$

the oxidized form is pale blue and the reduced is deep red. A solution containing $Fe(phen)_3^{2+}$ constitutes the indicator ferroin, widely used in oxidation-reduction titrations in volumetric analysis.

Deduce, from the following data, the acidity range over which this indicator would be suitable for titrimetry with vanadate(V) solutions:

$Fe(phen)_3^{3+}/Fe(phen)_3^{2+}$

H_2SO_4 (mol dm^{-3})	0.10	1.0	2.0	4.0	6.0	
E^0/V	0.91	1.01	1.06	1.14	1.23	

Vanadium(V)/Vanadium(IV)

H_2SO_4 (mol dm^{-3})	0.05	1.0	2.0	4.0	6.0	8.0
E^0/V	1.1	1.06	1.03	0.96	0.89	0.76

3. Isomerism in metal chelates

Werner's classical research on coordination compounds established the occurrence of several different types of isomerism — the existence of two or more compounds of the same empirical formula but with different arrangements of atoms within the molecule. The number of isomers of a given type is dependent on the coordination number and the preferred stereochemistry of the metal and also on the nature of the ligand, for example, whether it be unidentate or chelating, symmetrical or unsymmetrical. By isolation and identification of the isomers formed, Werner was able to determine the geometry of certain complexes of two particular metals, namely the octahedral shape of cobalt(III) complexes and the square-planar arrangement of ligands in platinum(II) complexes. His work was subsequently fully confirmed by definitive X-ray analyses and now for numerous coordination compounds the stereochemistry around the metal atom is known as a result of similar structural analyses.

Optical isomerism

This type of isomerism is found in coordination chemistry whenever a molecular structure is such that two forms, related as an object and its mirror-image, exist. The isomers are distinguishable by their effect on plane-polarized light. One isomer, the 'object', rotates the plane of polarization in one direction, whilst the other, the 'mirror-image', rotates it in the opposite sense. The two optically active isomers are called enantiomers and they are described as being enantiomorphous with each other.

In some instances of optical activity in inorganic compounds this is caused by a dissymmetric arrangement of molecules or ions in a crystal. Examples are quartz and crystals of $NiSO_4.6H_2O$. In coordination compounds, the ability to rotate polarized light originates in dissymmetry on a molecular scale, that is from the shape of the individual molecules or complex ions.

A dissymmetric molecule or ion lacks several kinds of symmetry elements: these include a centre of symmetry, a plane of symmetry, and a four-fold rotary inversion axis. As we shall see, a dissymmetric structure does have some symmetry elements; namely certain rotational axes.†
Many organic compounds show optical activity because they have an asymmetric carbon atom and have no element of symmetry except a one-fold rotation axis. Similarly, inorganic compounds which are asymmetric in the same sense are also expected to be optically active. In practice, dissym-

†See also Wormald: *Diffraction methods* (OCS 10) and Bishop (1973): *Group theory and chemistry*. Clarendon Press, Oxford.

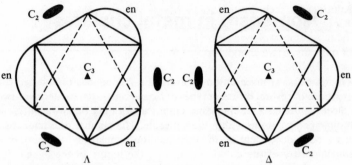

FIG. 3.1. Enantiomers of the tris(ethylenediamine)cobalt(III) ion.

metry, not asymmetry, is the cause of optical isomerism in metal complexes. Thus although a tetrahedral metal atom linked to four different unidentate ligands is the analogue of an organic compound containing an asymmetric carbon atom, no complex of this type has yet been resolved into optical isomers. Presumably separation has not been achieved because the isomers are so kinetically labile that rapid interconversion of one form to the other takes place.

As an example of a metal chelate which exists in enantiomorphous forms we may cite the tris(ethylenediamine)cobalt(III) ion. The two configurations Λ and Δ are shown in Fig. 3.1. If the chelate rings are assumed to be planar, this dissymmetric structure possesses one three-fold axis (C_3) and three two-fold axes (C_2) perpendicular to this. The enantiomers are related as object and mirror-image and neither can be superimposed on the other. This distinguishing feature can be best appreciated by construction of appropriate molecular models of the two forms.

So far we have discussed dissymmetry arising from the distribution of chelate rings about the central metal ion. Another cause of dissymmetry comes from the conformations of the chelate rings themselves. If these are puckered, their non-polarity can introduce dissymmetry. Also dissymmetry can be associated with the chelation of an optically active ligand.

The most common means of resolving complex cations or anions is by reaction with an optically active compound. For example, a solution containing racemic $(\pm)\text{Co(en)}_3\text{Cl}_3$ is resolved by reaction with barium $(+)$-tartrate. This results in the formation of a pair of diastereoisomers, $(+)\text{Co(en)}_3\text{Cl} (+)$-tartrate and $(-)\text{Co(en)}_3\text{Cl} (+)$-tartrate.† These are not enantiomorphous and can be separated by virtue of their different solubilities in water for on cooling a hot solution containing both diastereoisomers, $(+)\text{Co(en)}_3\text{Cl} (+)$-tartrate crystallizes out and the other isomer stays in solution. The optically active tartrate ion is removed by treatment

---

†The prefixes $(+)$ and $(-)$ refer to the signs of rotation at the sodium-D line wavelength, 589 nm.

of solutions of each diastereoisomer with sodium iodide. In this way, $(+)Co(en)_3I_3.H_2O$ and $(-)Co(en)_3I_3.H_2O$ may be separately prepared.

The resolution of dissymmetric anions can be achieved by salt formation with optically active alkaloids such as strychnine, brucine, or cinchonine. Resolved complex ions themselves can also be used. For example, $[(+)Ni(phen)_3]^{2+}$ can be employed for the formation of diastereoisomers of $[Co(C_2O_4)_3]^{3-}$. As before, the diastereoisomers differ enough in their physical properties for separation to be possible and then resolution is completed by removal of the resolving agent.

Geometrical isomerism

Geometrical isomerism represents different spatial distributions of a given set of atoms or groups about a central atom. One of the best-known examples is that of the two isomers of diamminedichloroplatinum(II). The square-planar distribution of bonds from platinum allows alternative arrangements of the pairs of similar ligands, either on the same side (*cis*) or on opposite sides (*trans*) of the metal atom.

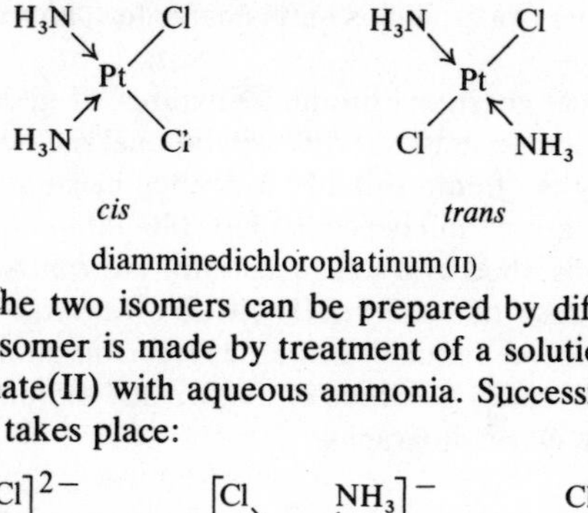

cis   *trans*

diamminedichloroplatinum (II)

In this case, the two isomers can be prepared by different synthetic routes. The *cis*-isomer is made by treatment of a solution of potassium tetrachloroplatinate(II) with aqueous ammonia. Successive replacement of two chlorines takes place:

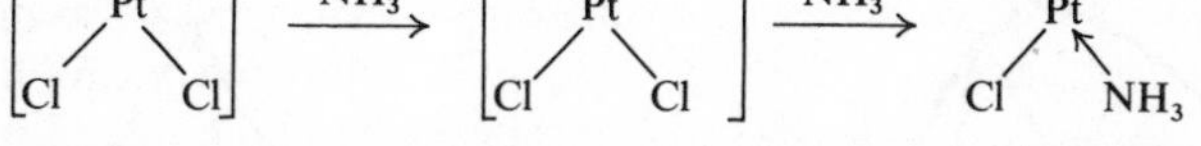

The least reactive chlorine in $[PtCl_3NH_3]^-$ is the one in the *trans*-position to ammonia and so one of the other chlorines is replaced, leading to the *cis*-isomer. From a study of the reactions of many platinum(II) complexes, it has been established that ligands differ in their ability to cause substitution in the position *trans* to themselves. This is known as the *trans* effect and evidently, from the above reaction route, the effect is greater for chloride ion than for ammonia.

The reaction leading to the formation of the *trans*-isomer also illustrates the *trans* effect. The tetraamineplatinum(II) ion $[Pt(NH_3)_4]^{2+}$ is reacted with excess chloride ion:

$$H_3N \diagdown \diagup NH_3 \xrightarrow{\ Cl^- \ } \quad H_3N \diagdown \diagup Cl \xrightarrow{\ Cl^- \ } \quad H_3N \diagdown \diagup Cl$$

The greater *trans*-directing influence of chloride ensures that the second chlorine substitutes in the *trans*-position to the first.

Reactions of platinum complexes of this kind with chelating ligands can be regarded as diagnostic of the stereochemistry of the isomer being studied. For example, *cis*-[Pt(NH₃)₂(NO₃)₂] reacts with oxalic acid to give a compound of empirical composition [Pt(NH₃)₂C₂O₄] whereas the *trans*-dinitrato complex yields one of composition [Pt(NH₃)₂(C₂O₄H)₂]. On the assumption that no change in the relative position of the two ammonia ligands occurs during these reactions, the replacement of two *cis*-nitrato groups by oxalate ion acting as a bidentate ligand takes place in one case whereas in the other, the *trans*-nitrato groups are replaced by two mono-protonated oxalate ligands acting in a unidentate manner. Because of its small size, a single oxalate ion is quite unable to span *trans*-positions in a metal complex.

Where chemical reactions produce mixtures of geometrical isomers instead of pure compounds, various separational methods are available. Chromatography is often a suitable technique because the two isomers differ in their polarities and hence in their solubilities and in the extent to which they are adsorbed by a solid substrate. The *trans*-isomer either has zero (as in the case of *trans*-Pt(NH₃)₂Cl₂) or a small dipole moment, whereas the *cis*-isomer usually has an appreciable moment. The *trans*-isomer is the less readily adsorbed of the two and therefore the first to be eluted in column chromatography.

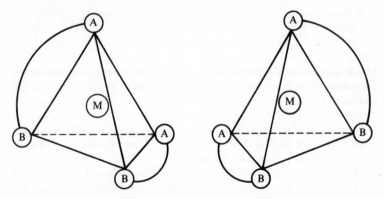

FIG. 3.2. Enantiomers of tetrahedral M(AB)₂ complexes.

Isomerism in complexes of bidentate ligands

Tetrahedral chelates

Optical isomerism is possible for all bis tetrahedral metal complexes of general type $M(AB)_2$, where AB represents an unsymmetrical chelating ligand. The two enantiomers are shown in Fig. 3.2. Relatively few have actually been resolved into optical isomers. Those which have include the bis-complex between 8-quinolinol-5-sulphonic acid and zinc ions.

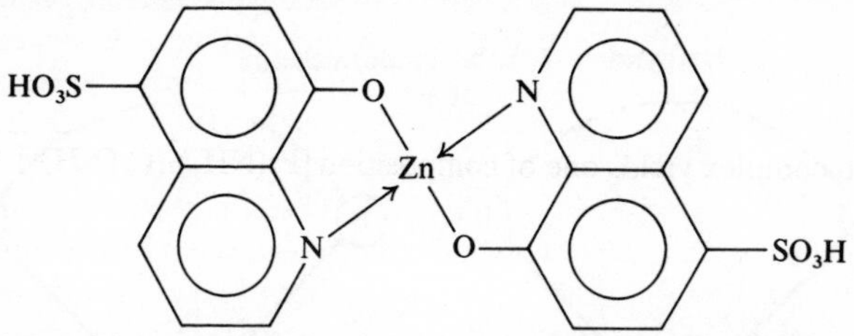

bis-complex between 8-quinolinol-5-sulphonic acid and zinc

Optical activity can result only if the bidentate ligand has an unsymmetrical structure. Thus a symmetrical ligand (AA) forming a planar chelate ring gives a complex ion $M(AA)_2^{n+}$ when coordinating to a tetrahedral metal atom and, because this complex has two planes of symmetry, each containing one of the chelate rings and bisecting the opposite triangular face of the tetrahedron, it cannot show optical isomerism.

Planar chelates

Optical activity is not normally possible in planar chelates because the arrangement of ligands usually results in the existence of one or more symmetry planes. For example, planar $M(AA)_2$ and $M(AB)_2$ complexes possess symmetry planes and are optically inactive.

When the chelating ligand is unsymmetrical, two geometrical isomers are possible (Fig. 3.3): these are of the *cis*- and *trans*-type discussed earlier. Usually only one form can be isolated, because this is much more stable than the other. For example, in bis(glycinato)copper(II), the two planar glycine rings are attached to copper in the *cis*-arrangement. In bis-(pyridine-2-carboximido)nickel(II), the ligands have a *trans*-planar configuration.

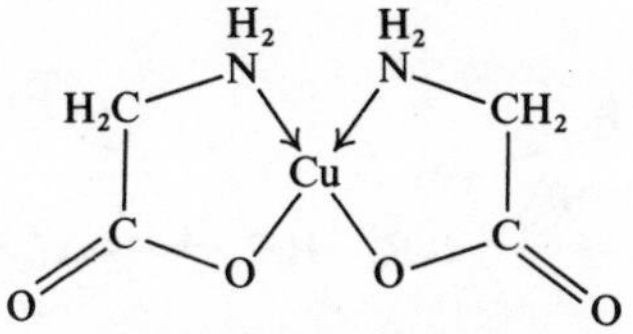

bis(glycinato)copper(II)

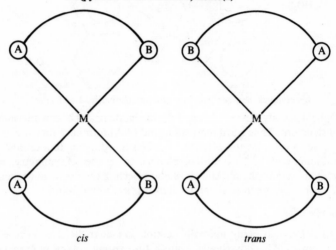

bis(pyridine-2-carboximido)nickel (II)

FIG. 3.3. Geometrical isomers of square-planar M(AB)₂ complexes.

Geometrical isomerism also occurs in a square-planar complex containing only one unsymmetrical chelating ligand, if the remaining coordination positions are occupied by different unidentate groups (C and D). The possible isomers are shown in Fig. 3.4. Here isomerism is intrinsically associated with the planarity of the structure, and does not depend on the presence of a chelate ring, because similar isomerism would be possible even if A, B, C, and D were all unidentate.

optical isomers of the isobutylenediamine-*meso*-stilbenediamineplatinum(II)ion

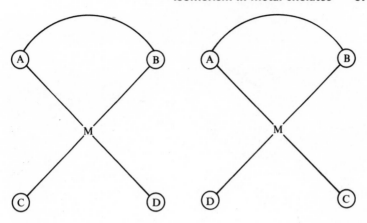

FIG. 3.4. Geometrical isomers of square-planar M(AB)CD complexes.

A classical demonstration of the square planarity of 4-coordinated platinum was achieved by Mills and Quibell in their resolution of the isobutylenediamine-meso-stilbenediamineplatinum(II) ion. The planar arrangement gives a structure which is resolvable into optical isomers because it possesses neither a plane nor a centre of symmetry. If the disposition of bonds around platinum were tetrahedral, there would be a plane of symmetry coincident with the plane of the platinum-isobutylene-diamine chelate ring.

Octahedral chelates

The occurrence of optical isomerism in tris-complexes of a symmetrical bidentate ligand with an octahedral ion has already been exemplified by $Co(en)_3^{3+}$. The resolution of this into optical isomers was a conclusive demonstration of Werner's theory of the octahedral disposition of the six bonds from cobalt. This is the only stereochemistry which leads to the existence of optical enantiomers. Alternative arrangements such as a planar hexagonal or a trigonal prismatic distribution of bonds must be discounted because optical isomerism cannot arise with either.

Possibilities of further isomerism exist if coordination of the ligand leads to non-planar chelate rings. A molecule like ethylenediamine can adopt several conformations in the free state because of internal rotation about the C–C bond. These are known as the *cis-*, *trans-*, and the two *gauche-*forms, designated λ and δ. These conformations of ethylenediamine as viewed along the C–C bond are shown in Fig. 3.5. The *trans-*conformation has the NH_2 groups on opposite sides of the C–C bond and so cannot act as a chelating ligand. If the *cis-*conformation were to coordinate with a metal,

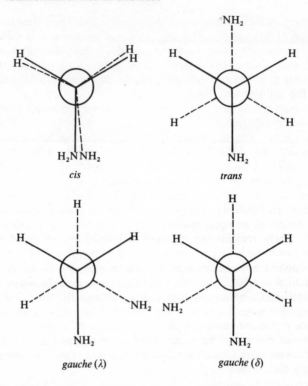

FIG. 3.5. Conformations of ethylenediamine as viewed along the C−C bond.

the chelate ring would be planar. If either *gauche*-conformation is involved, the chelate ring would be puckered. The λ and δ *gauche*-conformations are enantiomorphous but the energy of activation for the isomerization $\delta \rightarrow \lambda$ is too small for resolution of free ethylenediamine into these optical isomers to be possible. X-ray analysis of $[Co(en)_3]Cl_3.3H_2O$ indicates that all the ethylenediamine molecules are coordinated in a *gauche*-conformation. Because either the λ or δ conformation can be involved, there are eight possible configurations for a tris(bidentate) octahedral complex with non-planar rings:

$\Delta(\delta\delta\delta)$	$\Delta(\delta\delta\lambda)$	$\Delta(\delta\lambda\lambda)$	$\Delta(\lambda\lambda\lambda)$
1	2	3	4
$\Lambda(\lambda\lambda\lambda)$	$\Lambda(\lambda\lambda\delta)$	$\Lambda(\lambda\delta\delta)$	$\Lambda(\delta\delta\delta)$

The groups 1-4 represent pairs of enantiomorphous isomers. Conventional X-ray analysis can distinguish between 1, 2, 3, or 4, but not between pairs of optical isomers. In terms of their stereochemistries, the difference between $\Delta(\delta\delta\delta)$ and $\Delta(\lambda\lambda\lambda)$ is that in the first the direction of the central C−C bond in each ethylenediamine molecule is approximately para*lel* to the C_3 axis, whereas in the second the C−C bonds are slanted *ob*liquely relative to this. The two isomers have accordingly been labelled respectively the 'lel' and 'ob' forms.

The most stable form of Δ-Co(en)$_3^{3+}$ in the solid state is Δ ($\delta\delta\delta$). The sequence of stability for the four isomers is: $\Delta(\delta\delta\delta) > \Delta(\delta\delta\lambda) > \Delta(\delta\lambda\lambda) > \Delta$-($\lambda\lambda\lambda$). On the basis of differences in interligand repulsion effects, the energy difference between the 'lel' and 'ob' forms has been estimated as 7.5 kJ mol^{-1}. This value is too small for separation of any of the four isomers to be practicable at ordinary temperatures.

As we have already seen, enantiomers of a dissymmetric ion such as Co(en)$_3^{3+}$ are labelled $(+)$ or $(-)$ respectively depending on whether they show *dextro-* or an equal amount of *laevo*-rotation at a specified wavelength. There remains the question of relating the observed sign of rotation with the actual configuration of the enantiomer. By use of the technique of anomalous diffraction of X-rays, the absolute configuration of a number of metal complexes has been determined and it is now known that for [Co(en)$_3$]$^{3+}$ the configuration labelled Δ is that of the enantiometer which shows *dextro*-rotation at the sodium-D line wavelength. Correspondingly, Λ is the *laevo*-rotatory form.

We consider now the tris-complexes of an unsymmetrical bidentate ligand, AB. In this case, both optical and geometrical isomerism are possible. Geometrical isomerism exists because of the alternative arrangements possible of the unsymmetrical ligand molecules in coordination with the metal. These, illustrated in Fig. 3.6, are designated respectively *fac* (facial) and *mer* (meridianal). In the *fac*-isomer, the three donor A atoms

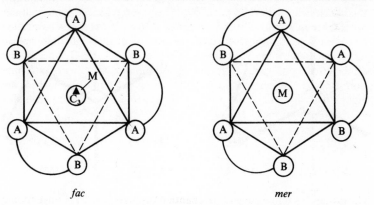

fac   *mer*

FIG. 3.6. Geometrical isomers of octahedral M(AB)$_3$ complexes.

$$H_2NH_2C \quad \quad H$$
$$C$$
$$H_2N \quad \quad CH_3$$
pn

are situated at the corners of one triangular face of the octahedron and the three donor B atoms are at the corners of another face parallel to the first. In the *mer*-isomer, the A atoms lie in a plane which is perpendicular to a similar plane containing B atoms. *Fac*- and *mer*-isomers can each exist in two enantiomeric forms.

The two geometrical isomers are clearly distinguished on grounds of symmetry. In the *fac*-isomer, all three ligands are in identical environments and the complex has a C_3 rotation axis. In the *mer*-form, the ligands are all in different environments relative to the donor atoms of the adjacent ligand the molecule has no element of symmetry. The uniqueness of each ligand is clearly seen if it is considered in relation to the donor atoms of the other two ligands in the *trans*-positions.

An example of a complex of the type $M(AB)_3$ is the tris-chelate formed between the enol form of benzoylacetone ($HBzC_6H_5COCH_2COCH_3$) and the cobalt(III) ion.

In the case of tris octahedral complexes of a ligand like 1,2-diamino-propane (pn), two geometrical isomers of the *fac*- and *mer*-type are again possible because the ligand is unsymmetrical in the same way as benzoyl-acetone. In addition, the ligand molecule contains an asymmetric carbon atom and is resolvable into optically active forms. When one of these enantiomers forms a chelate ring, it can coordinate in λ and δ *gauche*-configurations. These are illustrated in Fig. 3.7 for (+)-1,2-diaminoprop-

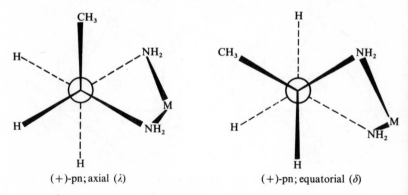

(+)-pn; axial (λ)   (+)-pn; equatorial (δ)

FIG. 3.7. λ and δ *gauche*-conformations of 1,2-diaminopropane.

ane. In the λ form, the methyl group lies in an axial position whereas in the δ form it is in an equatorial position. The chelate ring of the equatorial kind is estimated to be more stable than that of the axial type by about 9 kJ mol^{-1} because repulsions between the methyl and amino groups are less in the former than in the latter configuration. Thus the stable form of coordinated (+)-1,2-diaminopropane is δ and, correspondingly, that of (−)-1,2-diaminopropane is λ.

The complex ion $[M\{(+)pn\}]^{n+}$ can have two configurations, Δ and Λ, analogous to those of $[Co(en)_3]^{3+}$. Considering those forms with all three chelate rings of the same configuration, the most stable form is $\Delta(\delta\delta\delta)$ (lel); $\Lambda(\delta\delta\delta)$ (ob) is next in stability. These both contain equatorial methyl groups. The other two forms, $\Delta(\lambda\lambda\lambda)$ (ob) and $\Delta(\lambda\lambda\lambda)$ (lel), are relatively unstable because they have axial methyl groups.

In octahedral mixed complexes containing uni- and bidentate ligands, the number and type of isomers vary with the nature of the ligand. In bis-chelates of general formula $M(AA)_2X_2$, where X is a unidentate ligand, two geometrical isomers are possible, *cis* and *trans*, according to whether the chlorines are adjacent or opposite to each other in the coordination shell of the metal (Fig. 3.8). The *trans*-isomer has D_{2h} symmetry and

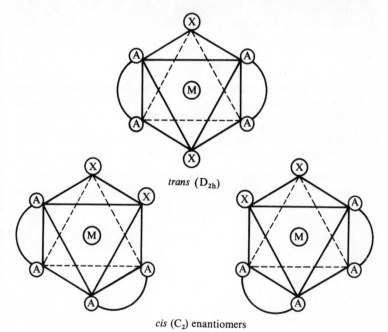

trans (D_{2h})

cis (C_2) enantiomers

FIG. 3.8. Isomers of octahedral $M(AA)_2X_2$ complexes.

optical isomerism is impossible, but the *cis* has C_2 symmetry and can exist in enantiomorphous forms.

Some of the best-known examples of geometrical isomerism among inorganic complexes are found in chelates of the type $M(AA)_2X_2$; for instance, the green *trans* (praseo) form and the violet *cis* (violeo) form of the $[Co(en)_2Cl_2]^+$ cation in its various salts. The configuration of the *cis*-isomer was established by Werner by its resolution into optically active forms and by chemical reaction with chelating ligands. For example, the two chlorines are replaced by a carbonato group in the reaction:

$$[Co(en)_2Cl_2]Cl + K_2CO_3 \rightarrow [Co(en)_2(CO_3)]Cl + 2KCl.$$

Providing no change of configuration around cobalt occurs during this reaction, the carbonato group can only replace ligands in a *cis*-relationship to one another.

In passing from $M(AA)_3$ to *cis*-$M(AA)_2X_2$, some symmetry elements are lost, the latter complex possessing only one C_2 axis compared with one C_3 and three C_2 axes in the former. In both, the presence of chelate rings confers disymmetry on the molecular structure and hence the property of optical isomerism. Analogous complexes in which the same donor groups are unidentate are more highly symmetrical and cannot exhibit optical activity.

PROBLEMS

3.1. In an octahedral complex $M(AB)_2X_2$, AB is an unsymmetrical bidentate ligand and X is unidentate. How many geometrical and optical isomers of this are theoretically possible?

3.2. How many isomers of $[Co(en)(pn)Cl_2]^+$ can exist? (pn = 1,2-diaminopropane; en = ethylenediamine).

3.3. Draw diagrams to represent the isomerism possible in complexes of general formula $M(AAB)_2$, where (AAB) is an unsymmetrical tridentate ligand.

4. Thermodynamic aspects

Enthalpy and entropy changes associated with chelation

For a system in equilibrium, the equilibrium constant K^T is related to the free energy change ΔG^0, by:

$$\Delta G^0 = -RT \ln K^T. \tag{4.1}$$

ΔG^0 is also related to the change in enthalpy ΔH^0 and the change in entropy ΔS^0, which accompany the reaction, by:

$$\Delta G^0 = \Delta H^0 - T\Delta S^0. \tag{4.2}$$

Thermodynamic data are available on many systems involving metal chelates. Some values of ΔG^0 and ΔS^0 for copper complexes with the N-containing ligands ammonia, ethylenediamine (en), 2-2'-2''-triamino-triethylamine (tren), and N,N'-di-(2-aminoethyl)ethylenediamine (diaen) are given in Table 4.1. From these we see that the free energy of formation is much smaller for copper-ammonia complexes (where no chelation can occur) than for copper-polyamine complexes (where chelate rings are formed). Values of $\Delta G^0(\beta_4)$ for copper-ammonia, $\Delta G^0(\beta_2)$ for copper-en and $\Delta G^0(\beta_1)$ for copper-diaen complexes are comparable because they all refer to essentially the same process, namely, removal of a certain number of water molecules from the hydrated copper(II) ion and the formation of four Cu$-$N bonds. These are directed, in accordance with the usual stereochemistry of 4-coordinate copper, towards the corners of a square. Cu(en)$_2^{2+}$ and Cu(diaen)$^{2+}$ differ from Cu(NH$_3$)$_4^{2+}$ in one major stereochemical aspect because they contain respectively two and three chelate rings (Fig. 4.1). It is stereochemically impossible for all four nitrogen donors in 'tren' to coordinate square planar copper and it is very probable that this ligand is tridentate in Cu(tren)$^{2+}$, the fourth coordination position being occupied by a water molecule.

The data in Table 4.1 suggest that metal chelates have an extra stability compared with analogous complexes of the same metal which contain no chelate rings. Schwarzenbach proposed the term 'chelate effect' to describe this extra stability and considered the origin and nature of this effect in relation to the changes in thermodynamic quantities accompanying complex formation.

According to eqn (4.1), a more stable complex is directly related to a more negative value of ΔG^0. From eqn (4.2), it is evident that such a change in ΔG^0 can be associated with either a more negative ΔH^0 value or a

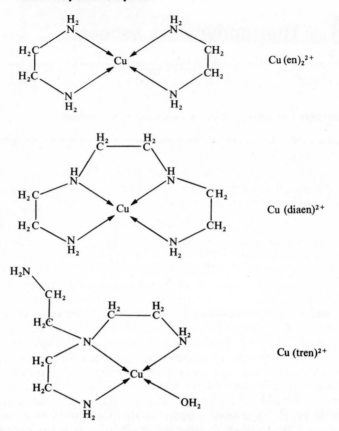

FIG. 4.1. The stereochemistry of some copper(II) complexes with nitrogen donor ligands.

more positive ΔS^0 value or with a combination of both these changes. Schwarzenbach concluded, on the assumption that enthalpy changes associated with the formation of metal-ligand bonds were the same whether the ligands were unidentate or chelating, that the special stability of chelates was primarily a result of the favourable entropy changes which accompany their formation.

Thus in the simple case of the formation of the complex ions $Cu(NH_3)_2^{2+}$ and $Cu(en)^{2+}$, the reactions can be represented by

(i) $Cu(aq)^{2+} + 2NH_3 \leftrightharpoons Cu(NH_3)_2^{2+} + 2H_2O$

(ii) $Cu(aq)^{2+} + n \leftrightharpoons Cu(en)^{2+} + 2H_2O.$

TABLE 4.1

(a) *Free energy changes for the formation of copper-ammine complexes*

Complex system	Free energy change (kJ mol^{-1})		Temperature	Medium
	ΔG^0	$\Delta G^{\ominus}$		
Cu^{2+}-NH$_3$	$\Delta G^0(\beta_2) = -44.85$	-64.79	298	2M NH$_4$ NO$_3$
	$\Delta G^0(\beta_4) = -74.48$	-114.22	298	2M NH$_4$ NO$_3$
Cu^{2+}-en	$\Delta G^0(\beta_1) = -59.83$	-69.62	293	Zero ionic strength
	$\Delta G^0(\beta_2) = -111.71$	-131.3		
Cu$_2$$^+$-diaen	$\Delta G^0(\beta_1) = -114.64$	-124.59	293	0.1M KCl
Cu^{2+}-tren	$\Delta G^0(\beta_1) = -107.52$	-117.48	293	0.1M KCl

ΔG^0 is the free energy change for one mole of product; $\Delta G^{\ominus}$ refers to the standard state of unit mole fraction.

(b) *Entropy changes for the formation of copper-ammine complexes*

Complex system	Entropy change (JK^{-1} mol^{-1})	
	ΔS^0	$\Delta S^{\ominus}$
Cu^{2+}-NH$_3$	$\Delta S^0(\beta_2) = -12.13$	$+54.64$
	$\Delta S^0(\beta_4) = -33.89$	$+99.66$
Cu^{2+}-en	$\Delta S^0(\beta_1) = +20.9$	$+54.31$
	$\Delta S^0(\beta_2) = +20.9$	$+87.70$
Cu^{2+}-diaen	$\Delta S^0(\beta_1) = +81.59$	$+114.97$
Cu^{2+}-tren	$\Delta S^0(\beta_1) = +75.3$	$+108.70$

In (*i*) ΔS^0 is negative; in (*ii*) ΔS^0 is positive (Table 4.1).

Assuming that the formation of each Cu$-$N bond in both cases takes place with the loss of one water molecule from the coordination shell of the aquated copper ion, we see that in (*i*) there are three independent species before and three after the reaction, whereas in (*ii*) there are two independent species before but three after the reaction. There is thus an increase in the number of species resulting from chelation which does not occur in (*i*). The greater the number of chelate rings which one ligand can form, the greater the increase in the number of species produced and hence the more favourable the change in translational entropy.

The association between an increase in the number of particles, that is the 'disorder' within a system, and positive entropy change is superficially attractive here and has often been made in the literature. McGlashan has stressed that such an association can strictly be assumed only in certain special cases and that it is incorrect, in the case of reactions in dilute aqueous solution, to try to interpret an entropy increase in such a simple quasi-geometrical manner as an increase in disorder.

Change in translational entropy is only one of a number of entropy changes accompanying chelation. There will also be changes in solvation entropy of the ions and ligands, a change in rotational entropy of the ligand, a decrease in entropy of the ligand resulting from the loss of internal rotation whenever ring closure takes place, and an entropy change arising from the ligand assuming the conformation necessary to coordinate with the metal ion.

When the complex is formed in aqueous solution by reaction between positive metal ions and negatively charged ligands, entropy increases are particularly favourable. The reactants are hydrated to extents which depend on their charge and size: the higher the charge, the greater the solvation, and this may extend well beyond the first coordination shell around the ion. As it is not so highly charged, the complex is much less extensively hydrated and so there is a net release of solvated water molecules when it is formed. The partial molal entropies of aqueous ions are more negative the higher their charge and the smaller their size (because both these changes increase the orientation of water molecules around the ion). The relatively larger complex ions have much higher entropies because of the charge neutralisation which accompanies their formation and the reduction in the extent to which they are hydrated. Therefore the reaction between a positive metal ion and a negative ligand is usually associated with a significant entropy increase.

In the formation of many metal chelates, the increase in stability is at least partly, and often chiefly, the consequence of more favourable enthalpy changes involved.

Thus, for

$$Cu(NH_3)_2^{2+} + en \rightleftharpoons Cu(en)^{2+} + 2NH_3, \quad \Delta H^0 = -5.44 \text{ kJ mol}^{-1}$$

and for

$$Cu(NH_3)_4^{2+} + 2en \rightleftharpoons Cu(en)_2^{2+} + 4NH_3, \quad \Delta H^0 = -21.3 \text{ kJ mol}^{-1}.$$

In both cases, the negative value for ΔH^0 means an enhancement of the stability of the metal chelate compared with the ammine complex.

Thermodynamic data for some metal complexes of EDTA and two of its homologues are summarized in Table 4.2. In the case of the magnesium, nickel, and calcium chelates, there is a decrease in stability as the number n of carbon atoms in the chain bridging the nitrogens is increased from 2 to 4.

TABLE 4.2

Stability constants, enthalpy changes, and entropy changes for complexes of EDTA and its higher homologues

Ligand	Mg^{2+}	Ca^{2+}	Ni^{2+}	Cu^{2+}
$(CH_2COO^-)_2N(CH_2)_2N(CH_2COO^-)_2 \lg K$	8.8	10.8	18.7	18.6
$\Delta H^0 (kJ\,mol^{-1})$	14.6	−27.6	−31.8	−34.3
$\Delta S^0 (J\,K^{-1}\,mol^{-1})$	213.2	112.9	246.6	242.4
$(CH_2COO^-)_2N(CH_2)_3N(CH_2COO^-)_2 \lg K$	6.3	7.4	18.3	19.1
$\lg K$	38.0	−7.1	−28.4	−32.2
ΔS^0	246.6	112.9	250.8	250.8
$(CH_2COO^-)_2N(CH_2)_4N(CH_2COO^-)_2 \lg K$	6.3	5.8	17.5	17.5
ΔH^0	35.5	3.76	−29.3	−27.2
ΔS^0	225.7	125.4	234.1	238.3

(From Martell (1967). The chelate effect. *Adv. Chem. Ser.* **62.**)

For the calcium complexes, the decrease in stability from $n = 2$ to $n = 3$ is solely due to the change in ΔH^0 to a less negative value. For the magnesium complexes, the net decrease in stability results from the adverse effect of the change in ΔH^0 which outweighs the increase, favourable to chelation, in ΔS^0.

For all four metal ions, ΔH^0 becomes more positive as n increases from 2 to 3, in other words as the number of atoms in the chelate ring increases from 5 to 6. Thus the smaller ligand, EDTA, forms the complexes with the more favourable enthalpy changes, and this is the major factor accounting for their greater stabilities, that is, why the five-membered is more stable than the six-membered ring. If we assume that, for the three ligands listed in Table 4.2, all donor groups can coordinate to the metal, then in solution the metal ion reacts with ligand species in which the distances between charged donor groups are different. As n increases, these groups must be brought together from greater distances to form the chelate; the forces of repulsion to be overcome are greater, and so the enthalpy of formation of the chelate becomes less negative.

The superior stability of a five- over a six-membered ring is a general feature of the chelates of saturated ligands. Further examples are given in Table 4.3, which summarizes thermodynamic data for the oxalate and malonate complexes of divalent manganese, cobalt, and nickel, formed according to the reaction:

$$M^{2+} + L^{2-} \rightleftharpoons ML$$

(L = oxalate or malonate). For each pair of complexes of the same metal, the ΔG^0 values are consistently lower in the case of the malonate complex. These lower values are clearly associated with the more positive values of

TABLE 4.3

Thermodynamic data for the 1:1 oxalate and malonate complexes of nickel(II), cobalt(II), and manganese(II) at 298 K.

	ΔH^0 (kJ mol^{-1})	ΔG^0 (kJ mol^{-1})	ΔS^0 (JK^{-1} mol^{-1})
Oxalate			
Ni	0.6	−29.5	101.3
Co	2.5	−27.4	100.0
Mn	5.9	−22.6	95.8
Malonate			
Ni	7.4	−23.4	103.8
Co	10.8	−21.5	107.9
Mn	14.8	−18.7	112.1

ΔH^0 because ΔS^0 shows very little difference within the set of six complexes. Again, the increase in stability in passing from Mn^{2+} to Ni^{2+} complexes of the same ligand results essentially from the enthalpy change becoming more favourable to chelation. The positive values for ΔS^0 illustrate again the increase in entropy which accompanies the coordination of charged ligands.

Another enthalpy effect arises from the ease or difficulty which the ligand finds in assuming the conformation most effective for chelation. Thus the enthalpies of formation of $Cu(diaen)^{2+}$ and $Cu(tren)^{2+}$ are respectively −90.17 and −85.35 kJ mol^{-1}. Although these are of comparable magnitude, the first refers to coordination by 4 donor nitrogen atoms whereas the second refers to coordination by 3. Thus the enthalpy of coordination per nitrogen atom is much greater for $Cu(tren)^{2+}$ than for $Cu(diaen)^{2+}$: this could be a consequence of the much more compact structure of the free 'tren' ligand so that three of its donor groups probably do not have to change conformation very much to give the copper complex. In contrast, 'diaen' is likely to exist in solution as an extended form which requires major conformational change before quadridentate coordination to a metal ion becomes possible.

The chelate effect

The reality of a 'chelate effect' has been queried and it has been argued that it is merely the consequence of an arbitrary asymmetry in the usual choice of standard states. Upon removal of this asymmetry, the effect largely disappears.

The basis of the criticism is that, for measurements of equilibrium constants in aqueous solution, the standard state taken for water is unit mole fraction (that is, the pure substance) whereas the standard states for all solutes are normally taken to be unit molar (or molal) concentrations.

For the general case of the interaction in aqueous solution of a metal ion M and a ligand L, the equilibrium can be represented by:

$$M(aq) + pL \rightleftharpoons ML_p + qH_2O \qquad (4.3)$$

(charges are omitted here and subsequently).

The equilibrium constant K may be defined as a concentration quotient:

$$\frac{[ML_p][H_2O]^q}{[M(aq)][L]^p} \text{, or an activity quotient, } \frac{\{ML_p\}\{\bar{H}_2O\}^q}{\{M(aq)\}\{L\}^p}.$$

'Concentration' (or 'activity') stability constants are derived from these quotients by assuming that the concentration (or activity) of water is unity, that is, its standard state is unit mole fraction. In contrast, the concentration (or activity) of each solute is expressed in $mol\,dm^{-3}$ or $mol\,kg^{-1}$. There is therefore an asymmetry of standard states inherent in the expressions used to define stability constants.

We now consider the derivation of relationships which can be used to calculate 'corrected' standard free energy and entropy changes (based on symmetric states of unit mole fraction for solutes and the solvent) from 'experimental' ΔG^0 and ΔS^0 values (based on standard states of unit molarity for solutes and unit mole fraction for the solvent).

The chemical potential $\mu(A)$ of a given species A, in a mixture of substances A, B, C ... etc., is given by:

$$\mu(A) = \mu(A)^{\oplus} + RT \ln x(A), \qquad (4.4)$$

where $x(A)$ is the mole fraction and $\mu(A)^{\oplus}$ the standard chemical potential of A. When $x(A) = 1$, $\mu(A) = \mu(A)^{\oplus}$, and $\mu^{\oplus}$ refers to the standard state of unit mole fraction.

For complex formation, according to eqn (4.3),

$$\mu(M(aq)) + p\mu(L) = \mu(ML_p) + q\mu\,(H_2O) \qquad (4.5)$$

Since

$$\Delta G^{\oplus} = \mu(ML_p)^{\oplus} + q\mu(H_2O)^{\oplus} - \mu(M)^{\oplus} - p\mu(L)^{\oplus} \qquad (4.6)$$

$$\Delta G^{\oplus} = -RT \ln \frac{x(ML_p) \cdot (x(H_2O))^q}{x(M(aq)) \cdot (x(L))^p} = -RT \ln K_x, \qquad (4.7)$$

K_x is the equilibrium constant expressed in mole fractions. In dilute aqueous solution, the mole fraction of water is unity and

$$K_x \simeq \frac{x(ML_p)}{x(M(aq)) \cdot (x(L)^p}.$$

In practice it is often more useful to express equilibrium and stability constants in terms of concentrations or molalities rather than mole fractions. Now the relationships involving chemical potentials are of the same form as before but refer to different standard states.

The chemical potential of A is given by:

$$\mu(A) = \mu(A)^c + RT \ln c(A) \tag{4.8}$$

where $c(A)$ is the concentration (mol dm^{-3}) and $\mu(A)^c$ is the chemical potential of A in a solution of unit concentration (1 mol dm^{-3}). The corresponding equation for change in free energy is

$$\Delta G^c = -RT \ln \frac{c(ML_p) \cdot (c(H_2O))^q}{c(M(aq)) \cdot (c(L))^p} = -RT \ln K_c. \tag{4.9}$$

Similarly, in terms of molality,

$$\mu(A) = \mu(A)^m + RT \ln m(A), \tag{4.10}$$

where $m(A)$ is the molality (mol kg^{-1}) and $\mu(A)^m$ is the chemical potential of A in a solution of unit molality (1 mol kg^{-1}). The standard state of A is defined as that of unit molality and

$$\Delta G^m = -RT \ln \frac{m(ML_p) \cdot m(H_2O)^q}{m(M(aq)) \cdot m(L)^p} = -RT \ln K_m. \tag{4.11}$$

Mole fraction and molarity in aqueous solution are related by:

$$x(A) = \frac{c(A)}{c(A) + c(B) + c(C) + \dots c(H_2O)}.$$

For low concentrations of all solutes, $x(A) \simeq c(A)/(H_2O) \simeq c(A)/55.5$ (because 1 dm^3 of water contains 55.5 moles).

Combination of eqns (4.4) and (4.8) gives

$$\mu(A)^{\ominus} = \mu(A)^c + RT \ln 55.5. \tag{4.12}$$

Substitution in eqn (4.6) of analogous expressions for $\mu(ML_p)^{\ominus}$ etc., gives:

$$\Delta G^{\ominus} = \mu(ML_p)^c + RT \ln 55.5 + q\mu(H_2O)^{\ominus} - \mu(M(aq))^c -$$

$$-RT \ln 55.5 - p\mu(L)^c - pRT \ln 55.5$$

$$= \mu(ML_p)^c + q\mu(H_2O)^{\ominus} - p\mu(L)^c - \mu(M(aq))^c - pRT \ln 55.5$$

$$= \Delta G^0 - pRT \ln 55.5 \tag{4.13}$$

where

$$\Delta G^0 = -RT \ln \frac{c(ML_p) \cdot (x(H_2O))^q}{c(M(aq)) \cdot (c(L))^p}$$

We note here that $\Delta G^{\ominus}$ values may be calculated from experimental ΔG^0 data using a correction involving p, the number of ligands which react with one metal ion. The significance of this is seen if we compare the respective standard free energies, $\Delta G_1^{\ominus}$ and $\Delta G_2^{\ominus}$ for the reactions: (1) of M(aq) with p molecules of a unidentate ligand L, and (2) of M(aq) with one molecule of a p-dentate ligand L_p.

From eqn (4.13),

$$\Delta G_2^{\ominus} - \Delta G_1^{\ominus} = (\Delta G_2^0 - \Delta G_1^0) + (p - 1)RT \ln 55.5. \tag{4.14}$$

This in fact represents the free energy change for the chelate reaction:

$$ML_p + (L_p) \rightleftharpoons M(L_p) + pL. \tag{4.15}$$

The chelate effect is significant if $(\Delta G_2^{\ominus} - \Delta G_1^{\ominus})$ has large negative values but not if this difference is small or zero. The values of ΔG given in Table 4.1(a) (p. 63) show that indeed *the chelate effect does largely disappear when unit mole fraction standard states are used throughout.* The consequence of the removal of asymmetry of standard states in this way is that for pure substances (that is, in the solid or molten states) chelates, appear to have no special stability compared with non-chelated complexes.

When the standard entropy change is modified in the same way, it is easily shown that

$$\Delta S^{\ominus} = \Delta S^0 + pR \ln 55.5 \tag{4.16}$$

and that the entropy change for (4.15) is

$$\Delta S_2^{\oplus} - \Delta S_1^{\oplus} = (\Delta S_2^0 - \Delta S_1^0) - (p - 1)R \ln 55.5 \quad (4.17)$$

($\Delta S^{\oplus}$ is the standard entropy change referred to unit mole fraction standard states; ΔS^0 is the entropy change corresponding to ΔG^0). The application to ΔS^0 values of corrections according to eqn (4.16) leads to the values of $\Delta S^{\oplus}$ given in Table 4.1(b). Inspection shows that $\Delta S^{\oplus}$ for $Cu(NH_3)_4^{2+}$ is comparable in magnitude to that for $Cu(diaen)^{2+}$ and $Cu(tren)^{2+}$ and is actually somewhat more positive than that for $Cu\ en_2^{2+}$. In other words, when standard states of unit mole fraction are used, there is not an increase in entropy specifically associated with chelation.

As many of the applications of metal chelation involve reactions in solution, for example, in the analysis of metal ions, and in the interactions between metals and organic ligands in biological systems, it is then more logical to consider standard states other than unit mole fraction.

For example, the equations derived for free energy and entropy changes ($\Delta G'$ and $\Delta S'$) for reaction (4.15) based on standard states of unit molar concentration are:

$$\Delta G_2' - \Delta G_1' = (\Delta \dot{G}_2^0 - \Delta G_1^0) + (q_1 - q_2) RT \ln 55.5 \quad (4.18)$$

and

$$\Delta S_2' - \Delta S_1' = (\Delta S_2^0 - \Delta S_1^0) - (q_1 - q_2) R\ln 55.5, \quad (4.19)$$

where q_1 and q_2 are respectively the number of water molecules released when the complexing of M(aq) occurs with a unidentate and a p-dentate ligand.

If $(q_1 - q_2)$ is small or zero, then ΔG^0 and ΔS^0 values are little or not at all affected by the change in standard state and so a real chelate effect is indicated.

We may go further and consider standard states of much lower concentration than unit molar. Thus if the symmetrical standard states of 10^z mol dm^{-3} are employed (designated by the superscript $+$) the modified equations for ΔS^0 and ΔG^0 are:

$$\Delta G_2^+ - \Delta G_1^+ = (\Delta G_2^0 - \Delta G_1^0) + (p - 1) RT \ln 10^z, \quad (4.20)$$

$$\Delta S_2^+ - \Delta S_1^+ = (\Delta S_2^0 - \Delta S_1^0) - (p - 1) R\ln 10^z. \quad (4.21)$$

When z is negative, that is, for standard states below 1 mol dm^{-3}, there is a negative contribution to $(\Delta G_2^\circ - \Delta G_1^\circ)$ and a positive one to $(\Delta S_2^\circ - \Delta S_1^\circ)$ — an enhanced chelate effect.

Martell (1967) has pointed out that the fact that, for example, the entropy contribution resulting from the formation of metal chelate rings can be eliminated by changing the standard state of the substances undergoing reaction does not, of course, affect the experimental reality of the greater stabilities of metal chelate compounds in solution. Whatever the standard state chosen, the equilibrium constant itself shows an increasing disparity — as the solutions become more dilute — between the degrees of dissociation of metal chelates and analogous non-chelate complexes.

For example, in the system

$$Cu(NH_3)_4^{2+} + tren \rightleftharpoons Cu(tren)^{2+} + 4NH_3,$$

experimental observation shows that the equilibrium lies well to the right for all but extremely low tren concentrations. We express this fact by stating that $Cu(tren)^{2+}$ is more stable than $Cu(NH_3)_4^{2+}$.

This is supported by a comparison of stability constants for the two reactions:

$$Cu(aq)^{2+} + 4NH_3 \rightleftharpoons Cu(NH_3)_4^{2+} + q_1 H_2O \ (\beta_4 = 10^{12.7})$$

and

$$Cu(aq)^{2+} + tren \rightleftharpoons Cu(tren)^{2+} + q_2 H_2O \ \ (K_1 = 10^{18.8}).$$

These constants are calculated using concentration units of mol dm^{-3} and, in this case, $K_1 >> \beta_4$.

On the other hand, if concentration units of mol cm^{-3} were to be used, calculated values are $\beta_4 = 10^{24.7}$ and $K_1 = 10^{21.8}$. Now $\beta_4 >> K_1$ and apparently the chelate complex should be much less stable than the non-chelate.

The change in concentration units may be regarded as a change in solute state from 1 mol dm^{-3} to 1 mol cm^{-3}. The former is the more realistic standard to choose, being closer to the concentrations of solutions generally used in stability-constant determination. There is therefore no reason to doubt the existence of a real chelate effect in such circumstances.

PROBLEMS

4.1. Calculate the entropy changes for the following reactions at 298 K and comment on the results:

$$Zn^2 + 2NH_3 \rightleftharpoons Zn(NH_3)_2^{2+} : \Delta H^0 = -28.03 \text{ kJ mol}^{-1};$$
$$\lg K_1K_2 = 5.01$$

$$Zn^{2+} + en \rightleftharpoons Zn(en)^{2+} \quad : \Delta H^0 = -27.6 \text{ kJ mol}^{-1};$$
$$\lg K_1 = 6.15$$
$$(R = 8.314 \text{ J K}^{-1} \text{ mol}^{-1})$$

4.2. The stepwise enthalpies ΔH_n^0 and the stability constants K for the system $\overline{\text{Ni}^{2+}}$-en in aqueous solution at 298 K are as follows:

n	1	2	3
ΔH_n^0 (kJ mol^{-1})	−37.7	−38.4	−40.6
lg K_n	7.51	6.35	4.42

Calculate the standard free energy and entropy changes associated with the addition of each ligand.

4.3. Stability-constant data, corrected to zero ionic strength, and measured over the temperature range 283-313 K, for nickel(II) complexes of 1,3-diaminopropane are (temperatures (K) in brackets):

lg K_1 6.67 (283) 6.40 (293) 6.18 (303) 5.94 (313)

lg K_2 4.71 (283) 4.44 (293) 4.28 (303) 4.09 (313)

Using the relation

$$\frac{d \lg K}{dT} = \frac{\Delta H}{2.303RT^2},$$

calculate average enthalpy changes for the addition of the first and second ligands. Calculate also the entropy changes at 293 K and 313 K.

5. Aminopolycarboxylic acids

The aminopolycarboxylic acids, also known as 'complexones', are highly effective chelating ligands. They are essentially derived from the simple amino acid glycine, which contains one amino and one carboxylate group so situated in the molecule that a five-membered ring is formed upon chelation. Complexones contain several carboxyalkyl groups bound to one or more nitrogen atoms with the result that coordination of a single metal ion establishes several chelate rings.

Nitrilotriacetic acid, ethylenediaminetetraacetic acid, and other complexones

The most widely used complexones are NTA, nitrilotriacetic acid, and EDTA, ethylenediaminetetraacetic acid. The practical value of EDTA and NTA lies in their ability to form stable, water-soluble complexes with many metal ions. This property is shown towards even such ions as magnesium and calcium and is the basis of analytical methods for the determination of hardness in water caused by the salts of these two metals. This was one of the first applications of EDTA to be described, by Schwarzenbach in 1946, and since then numerous uses have been developed in analytical chemistry and in many other fields.

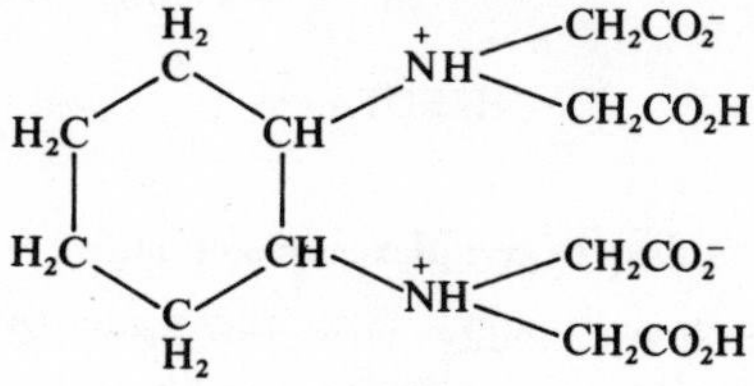

DCTA

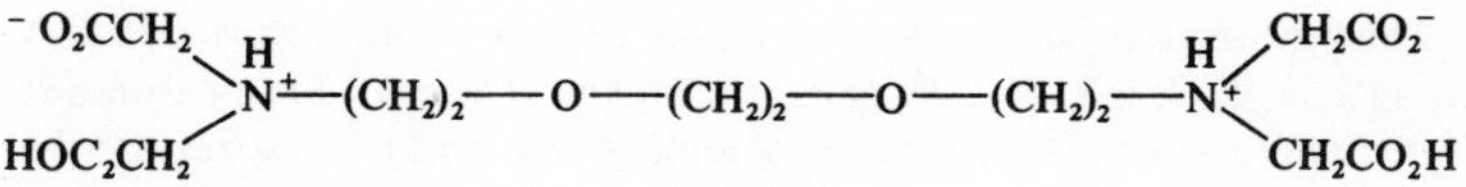

EGTA

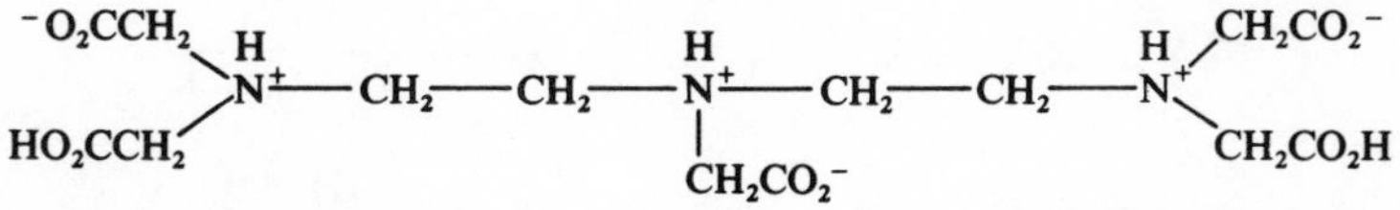

DTPA

Other complexones have been synthesized and their metal-complexing properties evaluated. Among the most useful of these are 1,2-diamino-cyclohexanetetraacetic acid (DCTA), ethyleneglycolbis(2-aminoethyl-ether)tetraacetic acid (EGTA), and diethylenetriaminepentaacetic acid (DTPA). These three all contain imidodiacetic acid groups linked in different ways and, as a result, the stabilities of their metal chelates show variations which can be exploited practically. Many other complexones are known in which the fundamental chelating unit of glycine is combined with other donor groups, for example, with phenolic groups in ethylene-diaminedi(o-hydroxyphenylacetic acid), EDDHA:

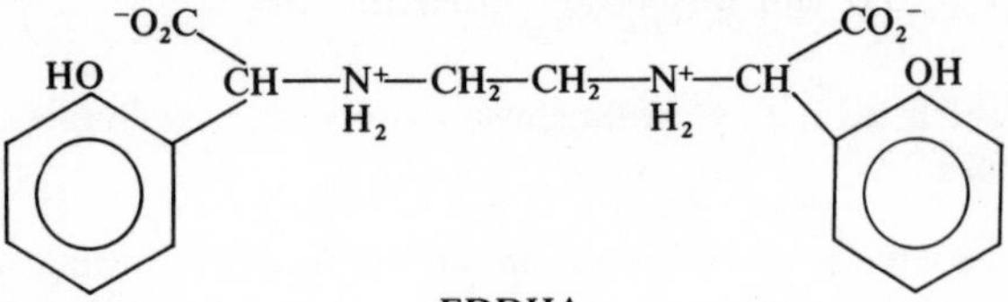

$$\text{HO} \quad \overset{^-O_2C}{\underset{}{CH}} - \overset{+}{\underset{H_2}{N}} - CH_2 - CH_2 - \overset{+}{\underset{H_2}{N}} - CH \overset{CO_2^-}{\underset{}{}} \quad OH$$

EDDHA

or with an alcoholic OH group in hydroxyethylethylenediaminetriacetic acid, HEEDTA:

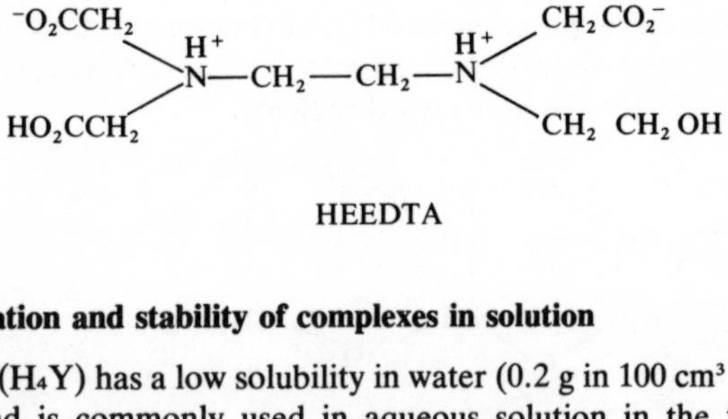

$$\overset{^-O_2CCH_2}{\underset{HO_2CCH_2}{}} \overset{H^+}{N} - CH_2 - CH_2 - \overset{H^+}{N} \overset{CH_2CO_2^-}{\underset{CH_2 \ CH_2 OH}{}}$$

HEEDTA

The formation and stability of complexes in solution

EDTA (H_4Y) has a low solubility in water (0.2 g in 100 cm^3 of water at 22 °C) and is commonly used in aqueous solution in the form of its disodium salt $Na_2H_2Y.H_2O$ (solubility = 10.8 g in 100 cm^3 at 22 °C).

The proton association constants for the various anionic forms of EDTA are $\lg k_4 = 2.07$, $\lg k_3 = 2.75$, $\lg k_2 = 6.24$, and $\lg k_1 = 10.34$ (measured at 20 °C in potassium nitrate solution of ionic strength 0.1). These refer to the equilibria:

$$H_4Y \rightleftharpoons H_3Y^- + H^+ \qquad\qquad k_4 = \frac{[H_4Y]}{[H_3Y^-][H^+]}$$

$$H_3Y \rightleftharpoons H_2Y^{2-} + H^- \qquad\qquad k_3 = \frac{[H_3Y^-]}{[H_2Y^{2-}][H^+]}$$

$$H_2Y^{2-} \rightleftharpoons HY^{3-} + H^+ \qquad k_2 = \frac{[H_2Y^{2-}]}{[HY^{3-}][H^+]}$$

$$HY^{3-} \rightleftharpoons Y^{4-} + H^+ \qquad k_1 = \frac{[HY^{3-}]}{[Y^{4-}][H^+]}$$

Fig. 5.1 shows how the relative amount of each EDTA species varies with the pH. H_4Y has a double betaine structure (p. 4), and the two protons which are lost first (below pH = 4) come from the two carboxyl groups. Proton dissociation from one of the nitrogens becomes significant above pH $\simeq$ 5.0 and dissociation from the second, above pH $\simeq$ 9.

In aqueous solutions of pH 3-5, EDTA exists mainly as H_2Y^{2-}. In these circumstances, it reacts with metal ions as follows to form complexes of 1:1 stoichiometry:

$$M^{n+} + H_2Y^{2-} \rightleftharpoons MY^{(n-4)+} + 2H^+.$$

In the pH region 7-10, the corresponding reaction is:

$$M^{n+} + HY^{3-} \rightleftharpoons MY^{(n-4)+} + H^+.$$

The stability constants $K(MY)$ for the complexes (MY) of various metals are given in Table 5.1. In the general case, these are defined by

$$K(MY) = \frac{[MY]}{[M][Y]}.$$

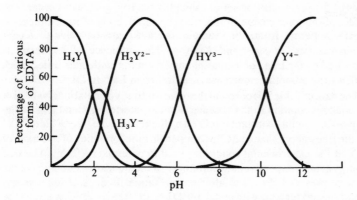

FIG. 5.1. Variation with pH of the different forms of EDTA.

TABLE 5.1

Stability constants of EDTA complexes†

Cation	lg K(MY)	Cation	lg K(MY)
Mg^{2+}	8.69*	La^{3+}	15.50
Ca^{2+}	10.96	Ce^{3+}	15.98
Sr^{2+}	8.63*	Pr^{3+}	16.40
Ba^{2+}	7.76*	Nd^{3+}	16.61
V^{2+}	12.7	Sm^{3+}	17.14
V^{3+}	25.9	Eu^{3+}	17.35
VO^{2+}	18.77	Gd^{3+}	17.37
Mn^{2+}	14.04	Tb^{3+}	17.93
Fe^{2+}	14.33*	Dy^{3+}	18.30
Fe^{3+}	25.1	Ho^{3+}	18.74
Co^{2+}	16.31	Er^{3+}	18.85
Ni^{2+}	18.62	Tm^{3+}	19.32
Cu^{2+}	18.80	Yb^{3+}	19.51
Zn^{2+}	16.50	Lu^{3+}	19.83
Hg^{2+}	21.8		

†lg K(MY) values at 293 K and $\mu = 0.1$ in KNO_3*
(or 0.1 in KCl).

Charges are omitted for the sake of brevity; the term [Y] represents the concentration of the fully deprotonated form of EDTA, Y^{4-}. The data demonstrate the versatility of EDTA as a chelating agent. Not only does it form very stable complexes with transition-metal and lanthanide ions but it is also able to chelate strongly with some metals, like the alkaline earths, which show generally much weaker complexing ability towards almost all other complexing agents. The stabilities of the complexes of the divalent ions of the first transition series exemplify the Irving-Williams order. The effect of the ionic charge of the metal on the stability of its complexes is clearly apparent from, for example, the much greater value of K_{VY^-} compared with $K_{VY^{2-}}$ and $K_{FeY^{2-}}$. The influence of ionic size is indicated by the trend of increasing stability in the lanthanide series which follows the monotonic decrease in radius from La^{3+} to Lu^{3+}.

The data in Table 5.1 represent the relative affinity of metal ions for EDTA in aqueous solution in the absence of other interfering metal ions or ligands. The stability constant is related by the relation $\Delta G^0 = -RT\ln K(MY)$ to the free energy change ΔG^0 accompanying the formation of one mole of MY at the experimental conditions used to determine K(MY). The free energy change depends on the difference of the standard free energy of the species present before and after complexation. If the solution contains other species reacting with metal or EDTA, the free energy change and the stability constant can vary considerably from their standard values.

It then becomes necessary to consider interaction of the metal ion with other ligands, of EDTA with other metal ions, and of hydrogen ions with EDTA or the metal complex. In order to do this, use may be made of the conditional constant for the species MY. The concept of a conditional constant originates in Schwarzenbach's work on the theoretical basis of the application of complexones in analysis (Schwarzenbach 1957) and was later developed by Ringbom (1963) for calculating the overall effect of a whole range of competing equilibrium reactions.

For MY, the conditional constant is defined by

$$K'(\text{MY}) = \frac{[\text{MY}]}{[\text{M}'][\text{Y}']}$$

Here and subsequently the primes indicate that the concentration terms [M'] and [Y'] include all forms of metal ion and ligand which are *not* complexed as MY.

For example, when zinc and EDTA ions are present together in an ammonia/ammonium-chloride buffer:

$$[\text{Zn}'] = [\text{Zn}^{2+}] + [\text{Zn(NH}_3)^{2+}] + \ldots [\text{Zn(NH}_3)_4{}^{2+}]$$
$$+ [\text{Zn(OH)}^+] + \ldots [\text{Zn(OH)}_4{}^{2-}]$$

and

$$[\text{Y}'] = [\text{Y}^{4-}] + [\text{HY}^{3-}] + [\text{H}_2\text{Y}^{2-}] + [\text{H}_3\text{Y}^-] + [\text{H}_4\text{Y}].$$

From these relationships, it is clear that the magnitude of the conditional constant for ZnY, $K'(\text{ZnY})$, depends on the concentrations of all species in solution involving zinc and EDTA.

For the general case of a second ligand A, which can also complex with M, then

$$[\text{M}'] = [\text{M}] + [\text{MA}] + \ldots [\text{MA}_n].$$

Schwarzenbach alpha coefficients

As a measure of the extent to which this type of competitive reaction takes place, Schwarzenbach utilized the *alpha coefficient*, defined by the equation:

$$\alpha(\text{M(A)}) = \frac{[\text{M}']}{[\text{M}]}$$

Expressing this in terms of the overall stability constants (β_1, β_2, etc.) of the complexes MA, etc., and the concentration [A] of uncomplexed ligand,

$$\alpha(\text{M(A)}) = 1 + \beta_1[\text{A}] + \beta_2[\text{A}]^2 + \ldots \beta_n[\text{A}]^n.$$

Similar coefficients can be defined for other competing reactions involving the metal ion.

In the case of EDTA (or any other ligand) the corresponding alpha coefficient is given by

$$\alpha(Y) = \frac{[Y']}{[Y]}.$$

It follows that conditional and stability constants are related by the equation,

$$K'(MY) = \frac{K(MY)}{\alpha(M(A))\alpha(Y)}.$$

Considering the protonated forms of Y,

$$[Y'] = [Y^{4-}] + k_1[H^+][Y^{4-}] + k_1k_2[H^+]^2[Y^{4-}] + \\ + k_1k_2k_3[H^+]^3[Y^{4-}] + k_1k_2k_3k_4[H^+]^4[Y^{4-}]$$

and so

$$\alpha(Y(H)) = 1 + k_1[H^+] = k_1k_2[H^+]^2 + k_1k_2k_3[H^+]^3 + \\ + k_1k_2k_3k_4[H^+]^4.$$

From the values of k_1-k_4 given earlier, $\alpha(Y)$ can be calculated as a function of the pH of the solution.
For example, at pH = 7,

$$\alpha(Y(H)) = 1 + 10^{10.34} \cdot 10^{-7} + 10^{16.58} \cdot 10^{-14} + 10^{19.33} \cdot 10^{-21} + \\ + 10^{21.4} \cdot 10^{-28}.$$

The first, fourth, and fifth terms are negligible compared with the others and so $\alpha(Y(H)) = 2.37 \times 10^3$.

Values of lg $\alpha(Y(H))$ are shown as a function of pH in Fig. 5.2. The higher the pH, the lower the value of $\alpha(Y(H))$. This rapidly approaches its limiting value of unity above pH 10.

Similarly, values for $\alpha(M(A))$ can be calculated for various ligands using stability-constant data for the appropriate complexes. Fig. 5.3 illustrates lg $\alpha(M(NH_3))$ as a function of lg[NH_3] for copper(II) and nickel(II) complexes with ammonia. The higher the concentration of ammonia and the greater the stability of the complexes, the greater the value of $\alpha(M(NH_3))$. Of the two metals, copper forms the more stable complexes and so the curve for $\alpha(Cu(NH_3))$ lies to the left of that for $\alpha(Ni(NH_3))$.

The data in Figs 5.2 and 5.3 can be used to calculate values for the conditional constant of the complex of either metal ion with EDTA.

For example, consider NiY in an ammonia-ammonium chloride buffer (containing 0.1 mol l^{-1} of NH_3) at pH 9.4. According to Fig. 5.2.

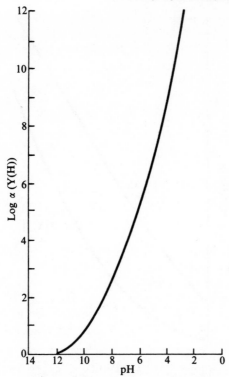

FIG. 5.2. Lg $\alpha(Y(H))$ as a function of pH.

lg $\alpha(Y(H)) = 1.0$, and according to Fig. 5.3, log $\alpha(Ni(NH_3)) = 4.2$. Therefore lg $K'(NiY) = $ lg $K(NiY) - $ lg $\alpha(Y(H)) - $ lg $\alpha(Ni(NH_3)) = 18.62 - 1.0 - 4.2 = 13.42$.

In many systems, several reactions can interfere with the complexation of a metal by EDTA, and so, for each of these, a corresponding alpha coefficient must be calculated in turn and its value used for the evaluation of the conditional constant. In addition to the normal complexes of stoichiometry MY considered above, EDTA also forms protonated complexes, MHY, hydroxo complexes, $MY(OH)_n$, and mixed complexes of the type MYX, where X is a unidentate ligand like H_2O, Br^-, or NO_2^-. If any of these are formed in a given system, the appropriate conditional constant defines its composition quantitatively.

The complexing properties of NTA (H_3X) have been extensively studied. The acid itself has a low water solubility (0.134 g in 100 cm³ at 5 °C) but its ammonium salt, $(NH_4)_2HX.H_2O$, and trisodium salt, Na_3X, are much more soluble.

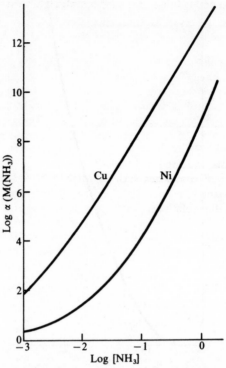

FIG. 5.3. Lg α (M(NH$_3$)) as a function of log [NH$_3$] (M=Cu or Ni).

NTA is a tribasic acid. The proton association constants for its various anionic forms are lg k_1 = 9.73, lg k_2 = 2.49, and lg k_3 = 1.89. These refer to the equilibria:

$$HX^{2-} \overset{k_1}{\rightleftharpoons} X^{3-} + H^+$$
$$H_2X^- \overset{k_2}{\rightleftharpoons} HX^{2-} + H^+$$
$$H_3X \overset{k_3}{\rightleftharpoons} H_2X^- + H^+.$$

The dissociation of protons from the two carboxyl groups occurs in an aqueous solution of quite low pH, but the betaine proton is not lost until the solution is appreciably alkaline. Over the range pH 4-8, the acid exists almost entirely in the form HX^{2-}. Reaction with metal ions proceeds thus:

$$M^{n+} + HX^{2-} \rightleftharpoons MX^{(n-3)+} + H^+.$$

In strongly alkaline conditions, the same complex is formed by the reaction

$$M^{n+} + X^{3-} \rightleftharpoons MX^{(n-3)+}.$$

The stoichiometry of the above complexes is the same as that of EDTA complexes. Because NTA acts at best as a quadridentate ligand, coordination of one X^{3-} ion can satisfy completely only those metals which normally show a coordination number of 4. It is not surprising therefore to find that some metals react with excess NTA with the formation of bis-complexes, $MX_2^{(n-6)+}$. The formation of comparable higher complexes, MY_2, with EDTA has not been observed.

In acidic media, the carboxyl groups can become protonated, leading to the production of complexes such as $MHX^{(n-2)+}$. The complex $MX^{(n-3)+}$ shows the property of a weak base in accepting a proton in this manner. In alkaline solutions, hydroxo complexes like $MX(OH)^{(n-4)+}$ can exist.

Stability-constant data for a selection of complexes MK are listed in Table 5.2. For each metal, the stability of its NTA complex is several orders of magnitude lower than that of its EDTA complex. This is in keeping with the lower chelating capability of NTA.

TABLE 5.2
*Stability constants
of NTA complexes*

Cation	$\lg K(MX)$
Mn^{2+}	7.44*
Fe^{2+}	8.83*
Co^{2+}	10.4
Ni^{2+}	11.54
Cu^{2+}	12.96
Zn^{2+}	10.66

†$\lg K(MX)$ values at 293 K
and $\mu = 0.1$ in KNO_3* (or 0.1
in KCl).

Structure of EDTA complexes

EDTA is capable of coordination as a sexadentate ligand and the stereo-chemistry of the cobalt(III) complex, $[CoY]^-$, in which EDTA behaves thus, has already been described. This is a rather special case in that the coordination number of the metal is fully satisfied by the ligand using all its donor capacity. It appears that octahedral coordination by the EDTA anion is possible only with cations of a relatively small size. With metal ions of greater size, constraints within the molecular structure of the ligand prevent this, and the metal complexed by EDTA may still remain accessible to other ligands.

For example, in the salt $Na_2[Mg(H_2O)Y]5H_2O$, although EDTA acts as a sexadentate ligand, the metal ion is actually 7-coordinate and its stereochemistry approximates to that of a monocapped trigonal prism.

Coordination numbers above 7 have been identified in some EDTA complexes. For example, 9-coordination exists in $[La(H_2O)_3Y]^{2-}$, and is believed to be present also in $[Ca(H_2O)_3Y]^{2-}$, and 10-coordination is found in a lanthanum complex containing protonated EDTA, $[La(H_2O)_4YH]$.

EDTA does not always fully utilize its donor capacity. For example, it is only quinquedentate in complexes like $[Ni(H_2O)YH_2]$ and $[Cu(H_2O)YH_2]$. One carboxylate group is not coordinated and octahedral coordination of the metal is completed instead by a water molecule. This lies in an equatorial position where it is *cis* to one nitrogen and *trans* to the other.

Other complexones

The complexing ability of a number of complexones analogous to EDTA has been assessed. These molecules are variations on a common structural theme, and the main motivation for their synthesis and study is to investigate how selectivity towards metals is altered by changes in the framework linking the glycine moieties, or, alternatively, by replacement of some of these by other chelating groups.

TABLE 5.3
*Stability constants
of DCTA complexes†*

Cation	lg $K(MY)$
Mg^{2+}	10.3
Ca^{2+}	12.1
Ba^{2+}	8.0
Mn^{2+}	16.78
Co^{2+}	18.92
Zn^{2+}	18.67
Cu^{2+}	21.3
La^{3+}	16.26
Lu^{3+}	21.51

†lg $K(MY)$ values at 20 °C
and $\mu = 0.1$ in KNO_3.

DCTA behaves very much like EDTA except that the stability constants of metal-DCTA complexes (Table 5.3) are generally between 10 and 10^3 times greater than those of their EDTA counterparts. Both DCTA and EDTA can form a maximum of 5 five-membered rings. The structural difference is that the cyclohexane ring in DCTA, although not directly concerned in chelation, considerably restricts the freedom of movement of the two nitrogen donors, compared with the situation in EDTA. This probably accounts for the relatively slow establishment of equilibrium in solution between a metal ion and DCTA but, nonetheless, the complexes formed are significantly more stable. There is no noticeable improvement of selectivity in the behaviour of DCTA towards metals.

Applications of complexones

The ability of complexones to form very stable, soluble chelates with so many metals has led to the development of applications for this class of chelating agents *inter alia* in industry (Chapter 8), in biological systems (Chapter 7), in analysis, and in plant nutrition.

Analysis

Reactions in aqueous solution between EDTA and metal ions occur very rapidly. They lead to the formation of complexes, of simple (1:1) stoichiometry, that are generally highly stable. Such characteristics mean that these reactions are ideally suited to form the basis of volumetric procedures for the analysis of metals. The chief disadvantage is the unselective nature of the complexone for it will react similarly with most, if not all, of the metals present in a mixture.

However, the reaction is influenced by the pH of the solution and some selectivity towards metals can be achieved by control of this. In solutions of pH 1-3, tri- and tetravalent metal ions which form very stable complexes can be titrated without interference from divalent cations. Between pH 5 and pH 6, most divalent cations, except those of the alkaline earths, are complexed. The alkaline earths, in accordance with the relatively low stabilities of their complexes, can only be titrated in alkaline media, at pH = 10 or above.

One of the first-developed volumetric applications of EDTA involved the determination, by titration with standard alkali, of the hydrogen ions released in the complexing reaction. This is not as straightforward as it might seem because the solution under examination will probably contain a mixture of various protonated forms of EDTA and the amount of hydrogen ions determined is not necessarily an accurate measure of the metal ion present. Although this method was originally used in the determination of calcium and magnesium, it has been superseded by much more generally applicable procedures in which changes in the concentration of free metal ion, rather than of hydrogen ion, are monitored directly.

As the solution containing metal ions is titrated with EDTA, the concentration of uncomplexed metal decreases gradually until, at the equivalence point (that is, when the molar ratio of EDTA added to metal salt present is exactly 1:1), this quantity shows a very sharp drop. The size of this change is dependent, of course, on experimental conditions such as the pH of the solution and the nature of other ligands present, but is usually at least several orders of magnitude. The decrease in metal ion concentration can be observed during the course of a titration if a compound called a *metallochromic indicator* is present. This is an organic dyestuff selected on the basis of its ability to form, at the experimental pH, an ion of one colour and a complex of a different colour with the metal being titrated. For ease of observation, it is important that these colours should be as different from one another as possible. In addition, the complex formed between metal ion and indicator must be less stable than that between metal ion and EDTA. This ensures that, so long as there is a deficiency of EDTA present, the colour of the metal-indicator complex predominates, but when the equivalence point is reached, the complexone removes virtually all the metal ion and so the solution takes on the characteristic colour of the ionic form of the indicator.

Metallochromic indicators are chelating agents which contain chromophoric groups. The requirement concerning the stability of their complexes relative to those of EDTA is met in general by the arrangement of functional groups within their molecular structure. This is such that fewer chelate rings are formed when the metal-indicator complex is produced than when the metal is complexed by EDTA. By their nature, the compounds also behave as acid-base indicators. For this reason, the pH of solutions in which they are used must be strictly controlled by buffers if the colour changes during the complexometric titration of a metal are to be optimal.

The first metallochromic indicators used were murexide:

murexide

and Erichrome Black T:

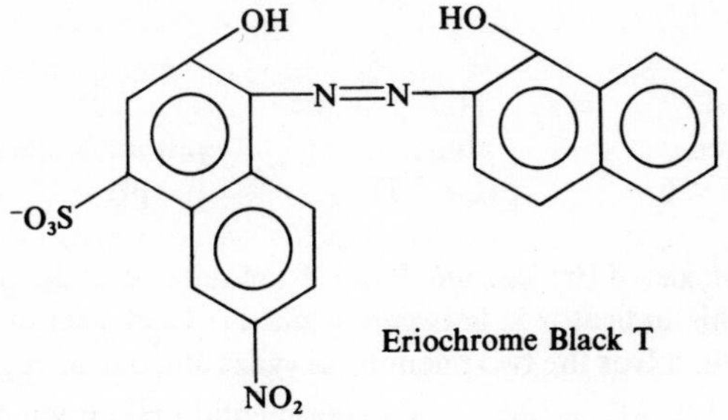

Eriochrome Black T

These were introduced by Schwarzenbach for the volumetric analysis of calcium and magnesium, respectively, using EDTA.

Murexide is the ammonium salt of purpuric acid and is represented above as its singly charged anion. In solutions up to pH = 9, this ion, reddish violet in colour, predominates. Above pH = 9, the solution becomes violet and above pH = 11, blue-violet, because of the progressive removal of protons from the imido groups. Murexide reacts with metal ions using its central nitrogen and the oxygens of the carbonyl groups as ligand atoms. With a few metal ions, notably those of calcium, nickel, and the rare earths, these complexes are sufficiently stable for use in volumetric analysis. The calcium complex is red and the colour change in an EDTA titration of calcium using this indicator is greatest at pH of 12 or above.

Eriochrome Black T is the first of a family of o-dihydroxyazo dyes which have been developed as metallochromic indicators. The sulphonic acid group loses its proton in quite acidic solutions and, below pH = 6, the

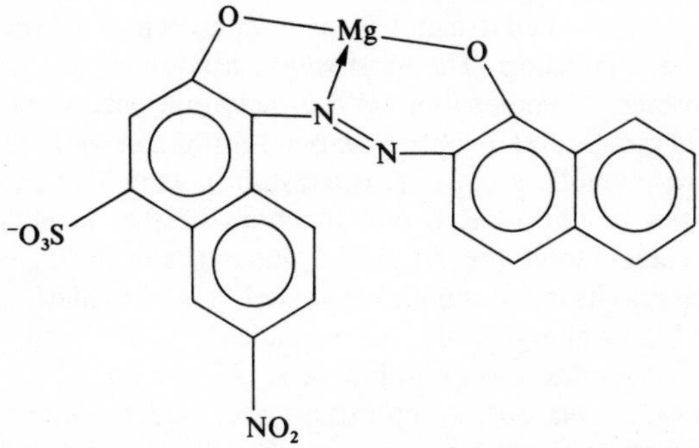

Eriochrome Black T chelated with magnesium

indicator exists as the singly charge anion (H_2D^-). This is red; successive deprotonation of the phenolic groups at higher pHs results in colour changes:

$$H_2D^- \quad \rightleftharpoons \quad HD^{2-} \quad \rightleftharpoons \quad D^{3-}$$

red	blue	yellowish orange
pH < 6	pH = 7-11	pH > 12

Metal complexes of Eriochrome Black T are red and so the pH range best suited for this indicator is between 7 and 11. Chelation of a metal like magnesium involves the two phenolic oxygens and one nitrogen of the azo group.

Calmagite contains the same chelating group of atoms as Eriochrome Black T and behaves in a very similar manner as a metallochromic indicator. Its chief advantage over the latter is its considerably greater stability in solution. Calmagite can be stored in solution for prolonged periods whereas Eriochrome Black T solutions decompose quite rapidly due to oxidation.

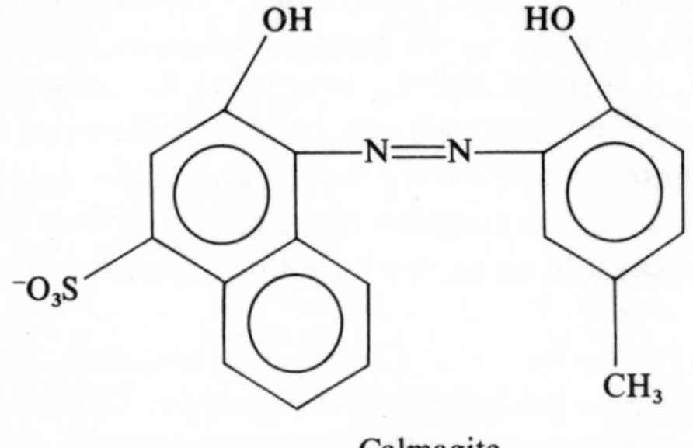

Calmagite

Another important group of indicators contains one or more imidodiacetic acid groups linked to a molecular structure which has the properties of an acid-base indicator. The most widely used indicator of this type is Xylenol Orange. Complexation with a metal ion occurs via the imidodiacetic acid groups and involves also bond-formation with the oxygens of the adjacent phenolic groups. This dyestuff has the acid-base indicator characteristics of Cresol Red and functions best as a metallochromic indicator in acidic solutions. At pH 3-5, the indicator itself gives a yellow solution whereas its metal complexes are red. It is particularly useful in the titration of highly charged ions, for example Zr^{4+}, Bi^{3+}, and Th^{4+}, which form EDTA complexes at quite low pHs.

In *gravimetric analysis*, complexones are used to prevent the precipitation of other metal ions which might otherwise interfere with the

Xylenol Orange

procedure for precipitation of a specified metal. By the formation of soluble complexes, the reactions of a metal ion can be masked and it is no longer precipitated by a reagent which characteristically gives an insoluble compound with the unmasked ion. For example, in the determination of calcium by precipitation as calcium oxalate, several cations interfere by forming insoluble hydroxides or oxalates. However, in the presence of EDTA in dilute acetic acid solution, interfering ions form stable, soluble complexes and are no longer precipitated, although calcium itself still reacts to form insoluble calcium oxalate.

Dimethylglyoxime is a selective reagent which has been used for many years for the determination of nickel and palladium. It can be used successfully for these metals only in conjunction with a masking agent when certain other transition metals are present. For example, iron and cobalt interfere with a nickel determination because they too can form sparingly soluble inner complexes with dimethylglyoxime. The traditional masking agent to prevent interference by iron is tartaric acid, because this forms a stable water-soluble iron complex. When iron and cobalt are present together, it has been established that they are still precipitated by dimethylglyoxime even in the presence of tartaric acid. In this instance, the complexone *N,N*-bis(hydroxyethyl)glycine is preferred to tartaric acid. With this ligand iron and cobalt form stable soluble complexes, which are unaffected by the addition of dimethylglyoxime, whereas nickel forms a weaker complex and can still be quantitatively precipitated from solution by dimethylglyoxime.

Plant nutrition

Plants need various nutrients for healthy growth and a deficiency in the supply of any one of these impairs the health of the plant. Essential nutrients include a number of metals such as iron, zinc, copper, manganese, boron, and molybdenum. The first four form stable complex anions with complexones and it is possible to use these to control the levels of the metals in soils or in nutrient solutions by supplying the right quantity of metal chelate within a specific pH range.

Iron in the +3 state is the most easily hydrolysed of these metals and is the one most commonly found to be deficient. Deficiency of iron results in many plant species in the disorder known as iron chlorosis. This is caused by a breakdown in the supply of iron required by the plant and is most strikingly demonstrated by the yellowing of the leaves. In calcareous soils, lime-induced chlorosis often develops. This is probably because the lime in the soil causes precipitation of iron in the form of compounds which have such a low solubility that plants are unable to absorb enough of the metal for their requirements.

The development of iron chlorosis has serious effects, for example, on the yield of fruit from citrus trees. For this reason, ways of combating chlorosis have been sought, and the most effective agents for doing so are synthetic iron chelates such as Fe(III)-EDTA and Fe(III)-EDDHA complexes. These act by supplying iron in a form that the tree or plant can readily assimilate. When the affected soil is treated with them the symptoms of chlorosis disappear and the fruit yield is dramatically increased.

In acid soils, the Fe(III)-EDTA chelate is effective for the restoration of health to chlorotic trees. In calcareous soils, this is only moderately successful. The chelate is not stable in solutions of pH 8 or above and iron tends to precipitate as insoluble hydroxide. Presumably in calcareous soils a similar process can occur to render the iron inaccessible to the tree. Alternatively, but with the same result, the iron may become irreversibly bound to clay constituents in the soil.

The iron(III) chelate of EDDHA is the most effective complex for the treatment of chlorosis in trees growing in calcareous soil. Because of the presence of phenolic groups, EDDHA has a very high affinity for iron(III) ions ($\lg K_1 = 33$) and the iron chelate shows no tendency to be converted to Fe(OH)₃ in alkaline solutions below pH 11.

Deficiencies of zinc, copper, and manganese in soils can be remedied in the same manner. Disorders due to lack of these metals are much less common than iron chlorosis because smaller amounts are required and also they are less susceptible than ferric iron to hydrolysis. One example of application on a commercial scale is the use of zinc-EDTA and zinc-HEEDTA in sprays to overcome zinc deficiency in cherry trees.

PROBLEMS

5.1. Calculate the concentrations of uncomplexed metal in a solution of calcium-EDTA (concentration of complex = 10^{-3} mol dm^{-3}) and in a solution of manganese-EDTA (concentration of complex

$= 10^{-2}$ mol dm^{-3}), taking the stability constants of the two complexes to be 10^{11} and 10^{14} respectively.

5.2. Using values given on p. 76 for the proton association constants of EDTA calculate values of $\alpha(H(Y))$ at pH 6.0 and pH 4.0. Check your values as closely as possible with those obtainable from Fig. 5.2.

5.3. In a volumetric procedure for the determination of hardness in water, 25 cm^3 of a water sample were titrated at pH 12 with EDTA solution (0.0096 mol dm^{-3}) using murexide as indicator. The titre (6.6 cm^3) represents the calcium content.

Another 25 cm^3 aliquot was titrated at pH 10 with EDTA solution of the same strength using Calmagite indicator. The titre (7.3 cm^3) represents the combined calcium and magnesium content.

Calculate the hardness caused separately by calcium and by magnesium ion. Express your answer in parts per million (p.p.m.) of each metal.

Atomic weights:
(Ca = 40.08; Mg = 24.3)

5.4. 25 cm^3 of a solution containing copper and zinc were titrated at pH 10 with EDTA solution (0.05 mol dm^{-3}) using murexide indicator. The titre (22.4 cm^3) represents the combined copper and zinc content.

Another 25 cm^3 aliquot of the copper and zinc solution was treated with 0.5 g of thiourea to complex the copper ions and then titrated at pH 6 with EDTA solution of the same strength using Xylenol Orange indicator. The titre (8.0 cm^3) represents the zinc content only.

Calculate, in g l^{-1}, the copper and zinc concentrations in the mixture.

Atomic weights:
(Cu = 63.5; Zn = 65.3)

5.5. Calculate the standard electrode potential for the redox reaction:
$$Co(EDTA)^- + e^- = Co(EDTA)^{2-},$$
given that $E^0 = 1.84$ V for Co(aq)$^{3+}$/Co(aq)$^{2+}$, the stability constants for Co(EDTA)$^-$ and Co(EDTA)$^{2-}$ are 10^{31} and $10^{16.3}$ respectively, $R = 8.314$ JK mol^{-1}, $T = 298$ K, and $F = 9.65 \times 10^4$ C mol^{-1}.

6. Keto-enol chelates

Tautomerism shown by β-diketones

The compounds acetylacetone(2,4-pentanedione) and similar β-diketones are of considerable interest because they show properties which are characteristic of a diketone structure and also behave in some respects as keto-enols. This class of compounds provides some of the best-known examples of the phenomenon of tautomerism — the existence of a single organic compound as a mixture of two forms differing structurally in the location of one hydrogen atom. Acetylacetone (Hacac) exists in tautomeric equilibrium with its enol form.

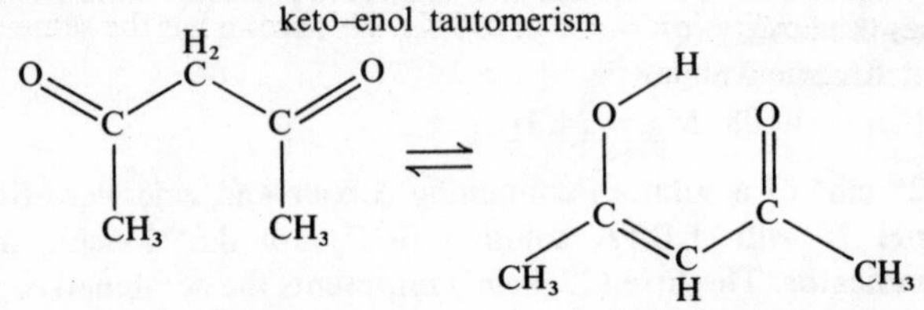

keto–enol tautomerism

keto form (diketo)   enol form (keto-enol)

The first metal derivatives of acetylacetone were described in 1887 and subsequent work has established the extent to which this ligand and other diketones are capable of coordination. By 1967, complexes with β-keto enols had been reported for all of the non-radioactive metallic or metalloidal elements in the periodic table, as well as for a number of the actinoids.

Acidity of enolic hydroxy group and metal chelation

In the keto-enol form, the ligand contains one acidic and one basic functional group. It coordinates as an anion through both oxygen atoms. Upon reaction with a metal ion, the positive charge on this is reduced by one unit for each ligand anion coordinated:

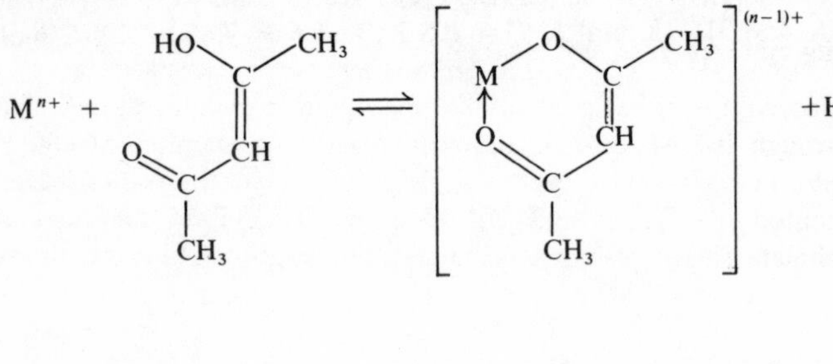

In the case where the coordination number of the metal is twice the number of positive charges on the ion, chelation with a keto-enol produces a neutral molecule:

$$M^{n+} + n(\text{Hacac}) \rightleftharpoons M(\text{acac})_n + n\text{H}^+.$$

Many neutral metal chelates have been made. They are usually sparingly soluble in water but dissolve in organic solvents, and are often quite volatile. It was the discovery of the unusual volatility of metal-containing derivatives of these ligands that led to the picturesque description of them as 'the ligands that give wings to the metals'.

The acidity of the enolic hydrogen of a β-keto-enol depends on the nature of the other groups in the molecule. Acetylacetone itself is a weak acid in aqueous solution ($pK = 8.93$ at 298 K). In trifluoroacetylacetone the presence of the strongly electronegative trifluoromethyl group increases the acidity ($pK = 6.3$ at 298 K). This group has the same effect in thenoyltrifluoroacetone ($pK = 6.3$ at 298 K).

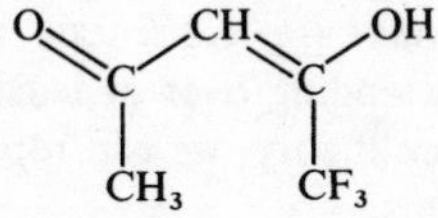

trifluoroacetylacetone

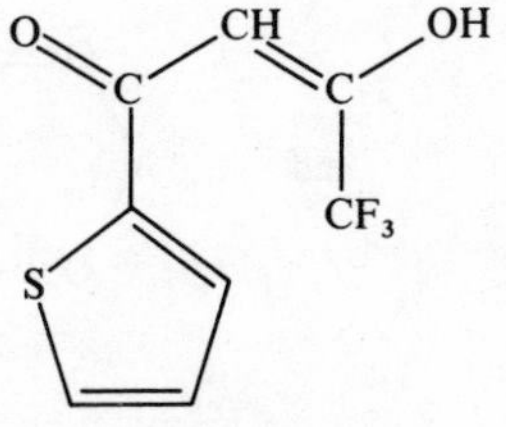

thenoyltrifluoroacetone

A practical consequence of the increased acidity of trifluoro derivatives is that chelation of a given metal ion occurs at a lower pH with these than with acetylacetone itself.

Bonding and structure of metal complexes

In most complexes with metals, acetylacetone coordinates as the enolate ion through its two oxygen atoms to form a six-membered ring. For example, in tris(acetylacetonato)iron(III), the iron atom is octahedrally coordinated by six ligand oxygens. X-ray crystallographic data show that each chelate ring is planar and that, within experimental error, the two

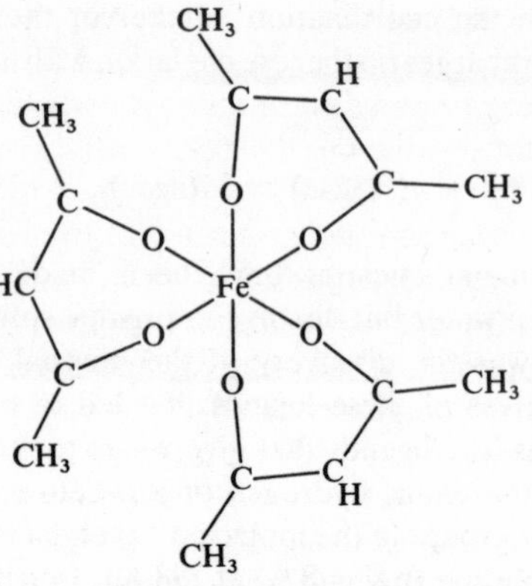

tris(acetylacetonato)iron(III)

C–C bond lengths are the same, and so are the two C–O bond lengths. Moreover, the C–C bond length of 137.5 pm is close to that (139 pm) in benzene. These observations are consistent with delocalization of electrons within the ring extending over at least the five atoms of the ligand. In terms of resonance theory, we can represent the bonding as a hybrid of two canonical forms.

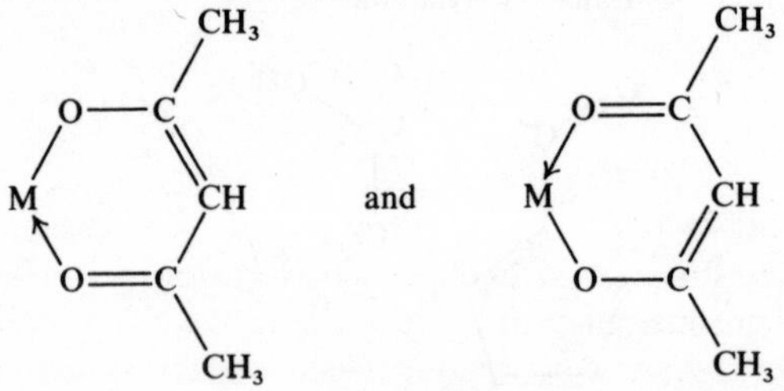

canonical forms of acetylacetone as ligand

This type of electronic structure is akin to benzenoid resonance and much attention has been paid in recent years to the question of the possible aromatic character of the chelate rings in acetylacetone complexes. One interesting aspect of this is the role of the metal in such a structure, namely whether or not it is able to participate in the π-system of the chelate ring by use of its d-orbitals. Experimental and theoretical studies of the electronic absorption spectra of transition-metal complexes of acetylacetone have led to the conclusion that d-orbitals of appropriate symmetry on the metal do indeed take part in π-bonding within the delocalized system. This involvement is believed to make a significant contribution to the stability of the chelate ring.

Support for the aromatic nature of the ring comes from chemical evidence that the hydrogen atom on the central carbon of the ligand can undergo various electrophilic substitutions; however, this aromatic-like behaviour need not involve the metal electrons as well as those of the ligand.

The concept of a 'ring current', associated with the delocalized electrons, has been widely invoked to interpret the n.m.r. spectra of aromatic molecules. In the case of the acetylacetone chelate ring containing several heteroatoms, it appears that such ring currents either are non-existent or they do not affect the properties to any significant extent.

A second type of acetylacetone complex which has been studied is that in which the ligand coordinates as the diketo tautomer via its two oxygens. This type is formed as, for example, a 1:1 adduct between acetylacetone and tin tetrachloride or titanium tetrachloride. The ligand behaves as a Lewis base and the metal chloride as a Lewis acid. The coordination number of the metal increases on adduct formation from 4 to 6.

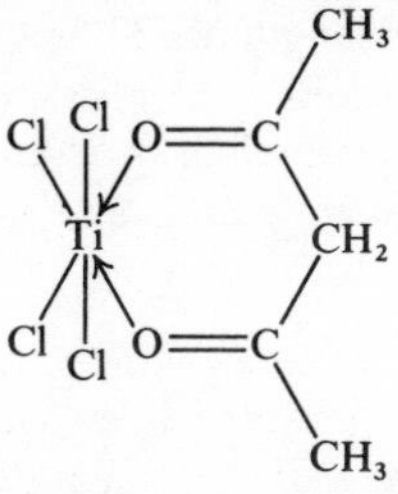

adduct between acetylacetone
and titanium tetrachloride

Another interesting feature of the structural chemistry of acetylacetone complexes is the existence of some compounds in polymeric forms. Polymerization appears to occur if the maximum coordination number of the metal is not attained in the monomer. By a process of sharing one or both oxygens of a ligand with two metal ions, the metal achieves a greater coordination number than in the monomer.

For example, in the monomeric form of bis(acetylacetonato)nickel(II), which is found in dilute solutions, in the vapour phase, or when the compound is isolated in an inert solid matrix, there is square-planar 4-coordination of the metal. In the crystalline solid, the complex is trimeric, $[Ni(acac)_2]_3$. In this form, each nickel atom attains 6-coordination in a slightly distorted octahedral environment by sharing oxygens (Fig. 6.1). The cobalt complex is a centrosymmetric tetramer (Fig. 6.2). This contains three distinct types of chelating ligands: those with both oxygens bound to only one cobalt atom, the terminal one; those with one oxygen serving as a bridge between two metal atoms; and those with both oxygens functioning

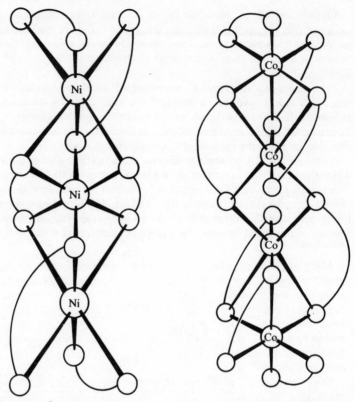

FIG. 6.1. The structure of [Ni(acac)₂]₃.   FIG. 6.2. The structure of [Co(acac)₂]₄.

as bridging atoms. As in the nickel trimer, the metal atoms are each coordinated by a distorted octahedron of oxygens. The distortion accounts for the observation that in neither polymer do the metal atoms lie in the planes of the chelate ring.

An interesting group of monomeric complexes containing 4-coordinated metal atoms is provided by those of an acetylacetone derivative, dipivaloylmethane(2,2,6,6-tetramethylheptane-3,5-dione). In these, polymerization does not occur, because the bulky t-butyl substituents prevent association of the type described above. Of those complexes with known structure, Cr(dpm)₃, Cu(dpm)₂, and Ni(dpm)₂ are planar monomers in the solid state and in non-coordinating solvents; Fe(dpm)₂, Co(dpm)₂, and Zn(dpm)₂ are tetrahedral (dpm represents the anion of the ligand dipivaloylmethane). This provides an importance sequence of compounds for establishing the stereochemical variation with metal ion

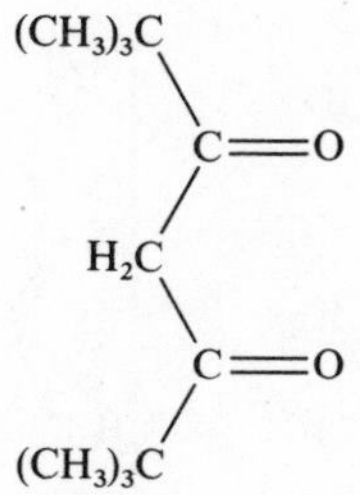

$$(CH_3)_3C$$
$$C{=}O$$
$$H_2C$$
$$C{=}O$$
$$(CH_3)_3C$$

dipivaloylmethane(2,2,6,6-
tetramethylheptane-3,5-dione)

where the ligand structure remains constant and imposes no apparent steric stabilization of either the planar or the tetrahedral configuration.

Reactions

Many organic reactions are catalysed by metal ions, and the catalytic action in most cases probably involves the intermediate formation of metal complexes with one or more of the reactants. In this context, a study of the reactivity of simple chelated ligands is particularly relevant. This provides basic information on how the reactivity of a ligand is altered by coordination to a metal, and should be helpful in understanding the catalytic functions of metal ions in much more complex systems such as those involving enzymes.

The β-keto-enol chelates are a class of complexes that has been very thoroughly studied from the point of view of the reactivity of the chelate rings. Chelates with certain metals are unsuitable because the chelate rings are broken easily by, for example, treatment with acids or the action of heat. However, Co(acac)$_3$, Cr(acac)$_3$, and Rh(acac)$_3$ are sufficiently stable to be of practical use.

In general, chelation of a ligand affects the rate at which it reacts but does not change the pathway or products of reaction. The reactions of the enol form of the ligand and metal acetylacetonate are similar because the ligand anion is in a similar environment in both cases. In one, it is coordinated to a proton; in the other to a positively charged metal ion. The enol form reacts more rapidly than the metal chelate but this is not necessarily true also for the neutral diketone.

Substitution under electrophilic conditions always involves replacement of the hydrogen atom on the central carbon atom $(C-3)$. Groups such as I, Br, Cl, SCN, SCl, NO$_2$, CH$_2$Cl, CHO, and CH$_2$N(CH$_3$)$_2$ have been substituted in one or more of the chelate rings in the tris chelate M(acac)$_3$. For example, bromination using N-bromosuccinimide proceeds thus:

The substitution occurs successively in each ring and an entering group appears to deactivate the remaining unsubstituted rings in the complex. For this reason, it is more difficult to make derivatives more highly substituted than the mono-substituted chelate.

Halogen atoms substituted in the 3-positions in metal acetylacetones are chemically quite inert. Thus the tri(3-bromo) derivative does not show the reactions typical of aryl halides, for example, it undergoes no reaction with magnesium or lithium in benzene solution. Nucleophilic displacement of bromine atoms by other groups is not possible. When this is attempted, either the chelate remains unaffected or the complex is destroyed completely.

Nitration with a mixture of nitric and sulphuric acids cannot be carried out because the chelate decomposes. However, copper(II) nitrate trihydrate in acetic anhydride is an effective nitrating agent and has been used successfully to make a tri(3-nitro)acetylacetonate. The nitro groups in such compounds are not reducible.

The unreactivity of substituents is attributable to two factors, the high electron density on C−3 and the steric shielding of this position by the flanking methyl groups. The steric effect is very pronounced in chelates such as Cr(dpm)$_3$. This undergoes chlorination or nitration extremely slowly because the large t-butyl groups on the ligand make it difficult for approaching groups to gain access to C−3.

In the case of ligands such as trifluoro- and hexafluoroacetylacetone, the presence of the electronegative CF$_3$ groups renders the central hydrogen atom inert to electrophilic substitution by reagents such as N-bromosuccinimide.

Applications of keto-enol chelates

In addition to the synthetic and structural aspects of the chemistry of keto-enol chelates, which have just been described, these compounds are of special interest to chemists because they have several important applications. The ligands themselves are among the most versatile of all chelating agents used for the solvent extraction of metal ions. Chelates of lanthanides such as europium and praseodymium have recently attracted a great deal of attention as shift reagents in the n.m.r. spectroscopy and certain europium complexes also exhibit the special property of laser action.

Solvent extraction

The keto-enol forms of ligands like acetylacetone, benzoylacetone, trifluoroacetylacetone, and thenoyltrifluoroacetone can react with many metals to form uncharged species which are extractable into an organic solvent immiscible with water.

Each of these ligands is a monobasic acid, represented generally as HR, which in aqueous solution is in equilibrium with hydrogen ion and the anion R^-, according to:

$$HR \rightleftharpoons H^+ + R^-.$$

The reaction between R^- and metal ion proceeds by the stepwise addition of ligands and the uncharged extractable complex is only one of a number of complexes in the equilibrium with each other. The equilibria involving the metal ion and its complexes may be written as:

$$M^{n+} + R^- \rightleftharpoons MR^{(n-1)+}$$
$$MR^{(n-1)+} + R^- \rightleftharpoons MR_2^{(n-2)+}$$

and so on up to

$$MR_{(n-1)}^+ + R^- \rightleftharpoons MR_n.$$

If the coordination number of the metal ion is not satisfied in the uncharged complex, additional ligands may be taken up to form anionic complexes like $MR_{(n+1)}^-$ etc.

When an aqueous solution containing these complexes is shaken with an immiscible solvent such as carbon tetrachloride or benzene (or alternatively when an aqueous solution of a salt of the metal is contacted with a solution of the ligand in the organic solvent) further equilibria involving partition of HR and MR_n between the two phases are established. The metal is distributed between the two solvents. Quantitatively, its distribution can be represented by the *distribution ratio D*, the ratio of the amount of metal in the organic phase to that in the aqueous phase. Assuming that MR_n is the only metal-containing species in the organic phase, that M forms no complexes in the aqueous phase other than those with R^-, and that the volumes of the two phases are equal, we find that

$$D = \frac{[MR_n]_O}{[M^{n+}] + [MR^{(n-1)+}] + \ldots [MR_n] + [MR_{(n+1)}^-]^+ \ldots},$$

where $[MR_n]_O$ is the concentration of the extractable complex in the organic phase. The concentration terms without subscript refer to the aqueous phase or, in terms of the overall concentration stability constants β of the various complexes,

$$D = \frac{[MR_n]_O}{M^{n+} (1 + \beta_1[R^-] + \dots \beta_n[R^-]^n + \beta_{(n+1)}[R^-]^{(n+1)} + \dots} .$$

When the concentration in the aqueous phase of all metal species other than M^{n+} is negligible,

$$D = \frac{[MR_n]}{[M^{n+}]} .$$

Using the relationships:

$$p(HR) = \frac{[HR]_O}{[HR]} ; \qquad p(MR_n) = \frac{[MR_n]_O}{[MR_n]} ;$$

$$K(HR) = \frac{[H^+][R^-]}{[HR]} ; \text{ and } \beta_n = \frac{[MR_n]}{[M^{n+}][R^-]^n}$$

and assuming that concentrations may be used in place of activities,

$$D = \frac{p(MR_n)(K(HR))^n \beta_n [HR]_O^n}{(p(HR))^n [H^+]^n} .$$

According to this equation the amount of metal extracted depends on the following factors: the partition coefficients of ligand and extractable complex, the acid strength of the ligand, the stability of MR_n, the pH of the aqueous phase, and the concentration of uncomplexed ligand in the organic phase at equilibrium. The applicability of this simplified expression for D depends on the correctness for a particular situation of the assumptions made in deriving it. If they are true, then the equation can be used to obtain information on the extractable complex. For example, a plot of lg D against lg $[HR]_O$ or pH should be a straight line of slope n, and this gives directly the value of n in MR_n.

D is related to E, the percentage of metal extracted, by

$$E = \frac{100D}{1 + D}.$$

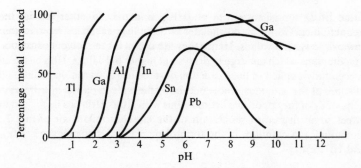

FIG. 6.3. The extraction of Al(III), Ga(III), In(III), Tl(III), Sn(II), and Pb(II) by acetylacetone in benzene (0.1 mol dm⁻³), as a function of pH (from Starý (1964). *The solvent extraction of metal chelates.* Pergamon Press, Oxford.)

Fig. 6.3 shows the variation of E with pH for the extraction of a number of Group III and IV metals by acetylacetone in benzene. The differences in extraction behaviour for the various metals reflect differences in the stability constants of the uncharged complexes and/or their partition coefficients, except where other factors become important, such as the formation to a significant extent of water-soluble complexes.

Thenoyltrifluoroacetone (Htta) and trifluoroacetylacetone (Htfa) have more acidic enolic groups than acetylacetone. This means that, other things being equal, the greater value of $K(HR)$ is reflected in a greater

TABLE 6.1

Distribution ratios (concentration in organic phase/concentration in aqueous phase) of actinide ions (M^{3+}) between Htta in toluene† and ammonium monochloro-acetate buffer‡ of pH = 3.4 at 298 K.

Metal	Distribution ratio
Ac	0.0001
Am	0.29
Cm	0.24
Bk	4.15
Cf	2.20
Es	1.60
Fm	2.58

†0.2 mol dm⁻³.

‡0.088 mol dm⁻³.

value of D for given values of $[HR]_0$ and pH. In other words, the extraction curves for metal chelates of these ligands tend to be displaced towards lower pH values. Hence they are useful for the solvent extraction of metal ions which undergo hydrolysis at higher pH values. Htta has been successfully applied to the extraction of some heavy transition metals and of some of the actinides. Table 6.1 shows the wide variation in extraction behaviour of the tripositive actinide ions with Htta. Htfa has been used to effect some degree of separation of the closely similar pair of metals, zirconium and hafnium, in the form of their tetrakis complexes $Zr(tfa)_4$ and $Hf(tfa)_4$.

Lanthanide shift reagents

Several paramagnetic complexes of the lanthanide form adducts with organic molecules carrying one or more donor atoms. Characteristically, these complexes cause substantial chemical shifts in the magnetic resonance spectrum for those protons in the organic molecule which are situated close to the site of coordination. There is much less line-width broadening

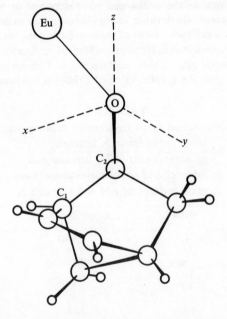

FIG. 6.4. Probable position of europium(III) ion in the adduct of $Eu(dpm)_3$ with *endo*-norborn-5-en-2-ol. The oxygen atom is at the origin of the internal coordinate system, the $O-C_2$ bond is placed along the negative z-axis, and the C_1 atom is placed in the xz-plane with the positive x-coordinate.

produced by a lanthanide complex than by an analogous first-row transition-metal complex, so the spectral resolution is appreciably increased, to produce in many instances a first-order spectrum. The paramagnetic lanthanide complexes are commonly known as 'shift reagents'. They are now widely used to obtain simplified ^{1}H n.m.r. spectra of organic molecules so complex that previously the application of n.m.r. spectroscopy could give little useful information because of the extensive overlap between resonance peaks typical of many-spin systems.

In 1969, the use of the europium complex Eu(dpm)$_3$(py)$_2$ as a shift reagent for cholesterol was first proposed. Subsequently, many applications have been developed for this and other shift reagents.

The most effective shift reagents of the β-keto-enol type are those in which bulky substituents are present, as in complexes of dipivaloylmethane. With such large ligands, the presence of strong interligand interactions limits the number of possible conformations of the shift reagent, perhaps even to a single one. Then the adduct of the reagent with an organic molecule is likely to be conformationally uniform, even though the latter by itself may show several conformations. The use of bulky substituents also lessens the affinity of the reagent for water, which, if absorbed, decreases its capacity to cause shifts.

Eu(dpm)$_3$ forms 1:1 association complexes with molecules like pyridine, n-propylamine, and neo-pentanol. In the crystalline state, the complex Eu(dpm)$_3$(py)$_2$ exists: X-ray analysis shows that there is square-antiprismatic coordination of the metal ion and that the pyridine molecules occupy apices of opposite square faces and are as far apart as possible. Such a complex can interact with a donor atom on an organic molecule either by expansion beyond 8 of the coordination shell of the europium atom or by replacement of a pyridine ligand. The most likely position of the metal ion in the adduct with *endo*-norbornenol is shown in Fig. 6.4.

Fluorine-substituted shift reagents such as Eu(fod)$_3$ have also been developed: 'fod' is an abbreviation for the coordinating 'heptafluoro-

fod

dimethyloctanedionato' anion. In its interactions with donor molecules, the coordination number of europium exceeds 6. For example, the species $Eu(fod)_3.2DMSO$ has been identified in CD_2Cl_2 solution.

Fluorescence spectra and laser action

When β-keto-enolate complexes of some trivalent lanthanide ions are irradiated with visible or near ultraviolet light, energy is absorbed by the organic chromophore and there may result the emission of narrow-line fluorescent radiation from the ion. This behaviour is shown, for example, by the benzoylacetone complexes of Eu^{3+}, Sm^{3+}, Tb^{3+}, and Dy^{3+}. Molecular phosphorescence and fluorescence also occur but these are negligible in comparison with the ionic fluorescence.

The phenomenon is the result of an intramolecular transfer of energy from the electronic levels of the organic ligand to localized levels within the 4f-shell of the ion. The small extent of phosphorescence from the chelate is indicative of an efficient transfer of energy intramolecularly. There are several stages in this transfer. First, absorption of energy produces an excited singlet from the ground singlet state in the ligand. Secondly, 'intersystem crossing' to a triplet state takes place by a radiationless process: competing with this are organic fluorescence and radiationless deactivation of the excited singlet as thermal energy. Thirdly, radiationless transfer of energy occurs from the triplet state to one or more low-lying 4f-levels in the ion; the requirement for effective transfer is that the triplet state be close in energy to or above the 4f-levels concerned. Fourthly, emission of energy follows if one of the 4f-levels excited by the transfer is a resonance level and transition to a lower 4f level can occur: this produces bright spectral lines characteristic of the individual lanthanide element.

A necessary condition for strong ionic fluorescence appears to be the existence of large energy gaps between the excited and lower 4f-levels of the lanthanide ion concerned. It is generally observed that competitive radiationless (quenching) processes become less probable whenever large quanta of energy must be dissipated, the transfer of energy rom ligand to ion is relatively more efficient and ionic fluorescence becomes more dominant.

If the radiation released in the final transition to a low-lying 4f-level is coherent, laser action results. For example, the europium(III)tris-(benzoylacetone) complex loses energy coherently at 16 329 cm^{-1} as a result of the internal transition from a $^5D_0$ to a $^7F_2$ level in the europium ion.

The fluorescence of europium chelates has been studied in view of its potential use in liquid lasers. Laser action is best achieved at low temperature in a medium of high viscosity and there are problems related to the dissipation in such a medium of the heat generated. For example, much of the work on the europium(III)tris(benzoylacetone) laser has been performed at $-150\,^\circ\mathrm{C}$ using a 3:1 mixture of ethanol and methanol. The performance characteristics are similar in many respects to solid-state lasers. The distinguishing property, however, is the very high value of absorption in the region of the pump band and consequent non-uniformity of excitation. The useful volume of the laser is thereby limited and this leads to a high threshold and a low efficiency. Efficient cooling of the laser cavity containing the chelate solution is necessary when using a methanol/ethanol mixture as solvent. However, similar fluorescence intensities have been reached using the same chelate at ambient temperatures in dimethyl-formamide or acetone. Evidently, there is considerable scope for the development of other solvent media to exploit more fully the laser action of this type of chelate.

PROBLEMS

6.1. The stability constants ($\lg K_1$) for the 1:1 complexes between some trivalent metal ions and acetylacetone are as follows:

Al^{3+}	Sc^{3+}	Y^{3+}	La^{3+}
8.6	8.0	6.4	5.1

Why do these values show a regular decrease on going from Al^{3+} to La^{3+}, and what does this trend indicate about the nature of the metal-ligand bonds?

6.2. Draw diagrams to illustrate the possible geometric isomers for the mixed ligand complex M(tfa)$_2$(acac). 'tfa' and 'acac' represent respectively the trifluoroacetylacetonate and acetylacetonate ions.

6.3. The chelate complex bis(hexafluoroacetylacetonato)copper(II), Cu(hfac)$_2$, reacts in concentrated solution in carbon tetrachloride with excess pyridine (C_5H_5N) to give an adduct which can be purified by sublimation *in vacuo* at 383 K. This adduct has the following percentage composition:

C: 37.8; H: 1.98; N: 4.40; Cu: 10.0.

Calculate the formula of the adduct.

Cu(acac)$_2$ forms an unstable adduct with pyridine and loses this readily when exposed to air. Why is the pyridine adduct with Cu(hfac)$_2$ more stable than that with Cu(acac)$_2$? (hfac = CF$_3$COCHCOCF$_3$. Atomic weights: C = 12.01; O = 16.00; H = 1.008; N = 14.01; F = 19.00; Cu = 63.54.)

6.4. One mole of tris(acetylacetonato)aluminium(III) reacts with 3 moles of dinitrogen tetraoxide in benzene solution. When the solvent is removed under vacuum, a yellow solid is formed of percentage composition: C: 44.1; H: 4.9; Al: 6.0; N: 7.3. Calculate the empirical formula of this compound and draw a diagram to illustrate its structure. (Atomic weight Al = 26.98.)

6.5. From an aqueous solution containing trace amounts of a radio-isotope of metal M, and 10 mg of inactive carrier, the complex $M(tta)_n$ is extractable by thenoyltrifluoroacetone ($0.10 \, mol \, dm^{-3}$) in benzene. Use the following data to determine the value of n:

pH	2.0	2.2	2.3	2.3	2.5	2.65	2.95
Per cent M extracted	4.0	13.1	22.5	37.6	55.1	77.0	90.3

7. Chelation in biological systems

The roles of metal ions

Metals are essential for the operation of many vital biological processes. Biologically important metals range from the 'bulk' metals, sodium, potassium, magnesium, and calcium, which together constitute more than 99 per cent of the total metal-ion content of a living organism, to iron and the 'trace' metals, manganese, cobalt, copper, molybdenum, and zinc, which are firmly bound within highly complex protein structures like the metalloenzymes and the haemoproteins.

The bulk metals are important in maintaining the structure of an organism, in controlling the functioning of cells, and in trigger and control mechanisms such as muscle contraction. The alkali metals, sodium and potassium, are present in biological fluids as their hydrated ions and it is the properties of these which are of primary importance in determining their functions. For example, sodium is the most abundant cation in fluids outside cells whereas potassium is the most abundant within cells. There are concentration differences for each metal between the interior and exterior of the cell, and these generate a potential difference across the cell wall. The existence of this potential permits nerve fibres to conduct impulses and hence ultimately are the cause of muscle contraction. The metal ions can diffuse freely into and out of cells and considerable attention has been paid to the mchanisms by which they are transported through the cell membrane. It seems likely that the immediate environment of the metal ion is altered appreciably during its transport through the membrane and quite possibly movement is facilitated by chelation processes involving naturally occurring ligands.

By using radioisotopes and ion-selective electrodes, it has been established that calcium and magnesium in biological fluids cannot move as freely as the simple hydrated ions. As we have seen already, these metals generally form stronger complexes than the alkali metals. It is not surprising, therefore, that at least part of the magnesium within a cell is in the form of nucleic acid complexes. These are involved in the transmission of nerve impulses, muscle contraction, and carbohydrate metabolism. Magnesium and calcium are also essential for the activation of certain enzymes. Thus some enzymes associated with the transfer or hydrolysis of phosphate groups require the presence of a magnesium ion. For example, magnesium must be present to activate the enzyme ATPase, which catalyses the hydrolysis of adenosine triphosphate to adenosine diphosphate and orthophosphate; and inorganic pyrophosphatase, which catalyses conversion of pyro- to orthophosphate. In both instances, it is likely that the magnesium ion is chelated by phosphato-group oxygens and this process serves to reduce the negative charge on the group and render it more susceptible to hydrolysis.

Calcium performs a wide variety of biological functions. In vertebrates it is required for the formation of bones and teeth, and is involved in a number of trigger and control mechanisms. The calcium content of a cell is normally very low, but a controlled influx of calcium ion initiates processes of muscle contraction. Although the nature of the calcium-containing species within the cell is not precisely known, it is almost certainly a complex rather than the hydrated ion. Calcium is found in some enzymes, for example, in α-amylase, which is responsible for the catalysis of the hydrolysis of glucosides.

Macrocyclic antibiotics

Certain naturally occurring antibiotics like nonactin, monactin, and valinomycin are multidentate ligands which show highly selective action in complexing metal ions. Nonactin and monactin are cyclic polyethers that contain 8 oxygens in a ring of 32 atoms.

These rings are quite flexible and they can expand or contract to accommodate ions of radii within a certain range. Thus, in dry acetone, nonactin can bind sodium, potassium, and caesium ions with approximately the same affinity. The stabilities of the metal complexes are greatly reduced when the solvent system is changed by the addition of water. The reductions are far greater for sodium and caesium ions than for potassium. In other words, nonactin selectively complexes potassium ion in the water-containing medium. The different extent to which these cations are hydrated in wet acetone is an important factor in the ion-selective

nonactin (R=H)   monactin (R=CH$_3$)

behaviour of nonactin. The alkali-metal ion must lose its water of hydration before it can be bound in the nonactin cavity. This process clearly occurs more easily with potassium than with other ions.

Valinomycin is the cyclic compound

$$\overline{\hspace{2cm}}$$
(D-valine-L-lactate-L-valine-D-hydroxyisovalerate)₃

and it contains a 36-membered ring which includes six ether oxygens and six peptide nitrogens. When valinomycin is dissolved in a solvent like n-hexane or octanol and placed in a conventional liquid-membrane electrode assembly, which is comparable in its selectivity with biological membranes of living cells, the electrode shows very high selectivity towards potassium compared with sodium ion. Valinomycin therefore is shown to enhance greatly the permeability of the membrane to potassium ion.

Its high selectivity makes the liquid-membrane electrode a very useful tool for analytical and biomedical measurements, where the determination of potassium ion in the presence of comparable or excess amounts of sodium ion is often required. Even more significant, perhaps, is the implication which this selective action holds concerning the mechanism of ion transport through lipid membranes. Thus it is possible that the valinomycin ligand acts as a 'carrier' molecule by complexing a cation within its cavity and that thus the ion, in the form of the valinomycin complex, moves through the membrane. The potentiometric selectivity of the valinomycin electrode is directly related to the relative stabilities of the alkali-metal complexes in solution and does not appear to be dependent on any process within the liquid membrane itself. Hence the transport of an ion through a biological membrane could similarly involve a carrier mechanism which depends primarily on the formation of complexes between the metal ion and valinomycin.

Metal complexes of nucleotides

The adenine nucleotides are some of the biologically most important compounds of relatively low molecular weight, for example, adenosine-5'-triphosphate (ATP), a carrier of chemical energy in a readily available form. Energy is released when ATP loses a phosphate group to an acceptor molecule and is converted to adenosine-5'-diphosphate (ADP).

Structurally, the phosphates of adenosine are multidentate chelating agents. The relationship between ATP, ADP, and AMP (adenosine monophosphate) is indicated in Fig. 7.1. These molecules are based on the adenine, in which a ribose group has been substituted in position 9. The phosphate chain is attached at the 5'-position of the ribose moiety.

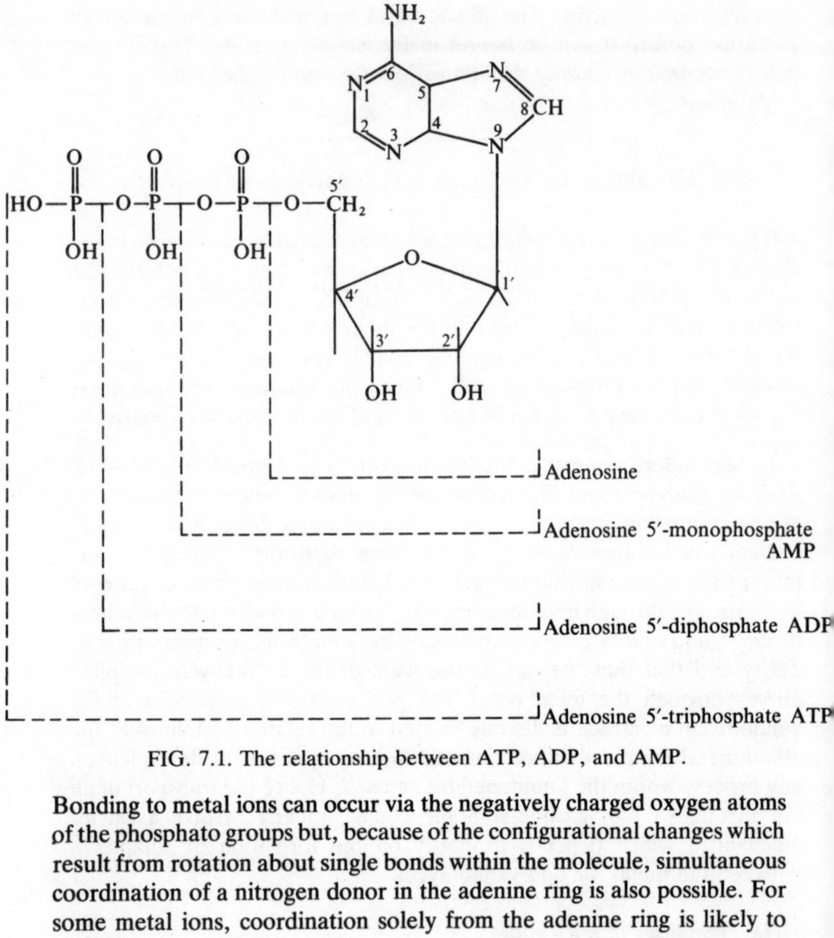

FIG. 7.1. The relationship between ATP, ADP, and AMP.

Bonding to metal ions can occur via the negatively charged oxygen atoms of the phosphato groups but, because of the configurational changes which result from rotation about single bonds within the molecule, simultaneous coordination of a nitrogen donor in the adenine ring is also possible. For some metal ions, coordination solely from the adenine ring is likely to occur.

In a solution containing a metal ion and an adenine nucleotide, equilibrium is established between the various forms of complex possible, 'metal-chain', 'metal-bridge', and 'metal-ring'. This equilibrium may be represented thus:

adenine-ribose-phosphate ⇌ adenine-ribose ⇌ adenine-ribose-phosphate

| | | |
M   M←phosphate   M

'metal-chain'   'metal-bridge'   'metal-ring'

TABLE 7.1

Stability constants for metal complexes of ATP
(a) *For* $ATP^{4-} + M^+ \rightleftharpoons (MATP)^{3-}$ *at* $\mu = 0.2$ *and* 25 °C

M	lg K(MATP)
Li^+	1.57
Na^+	0.96
K^+	1.00

(b) *For* $ATP^{4-} + M^{2+} \rightleftharpoons (MATP)^{2-}$ *at* $\mu = 0.1$ *and* 25 °C

M	lg K(MATP)
Mg^{2+}	4.22
Ca^{2+}	3.97
Sr^{2+}	3.54
Ba^{2+}	3.29
Mn^{2+}	4.78
Co^{2+}	4.66
Ni^{2+}	5.02
Cu^{2+}	6.13
Zn^{2+}	4.85

For a particular nucleotide, the proportion of each species varies according to the experimental conditions and the nature of the metal. For example, a metal ion which forms predominately electrostatic bonds, like Mg^{2+} or Ca^{2+}, tends to be virtually all in the form of a 'metal-chain' complex. This appears to be true also for chelating with much more complex species such as DNA and RNA. Metals like Cu^{2+} and Zn^{2+} which bond strongly to nitrogen as well as to oxygen donors tend to form 'bridge' complexes whereas an ion such as Ag^+, complexing strongly with nitrogen but not with oxygen, forms a 'ring' complex.

The stability constants of some metal-ATP complexes are listed in Table 7.1. These follow the sequence transition metal > alkaline-earth metal >> alkali metal. The predominance of metal-chain bonding is inferred from stability-constant studies of complexes of other nucleotides: these show that the stabilities of complexes between adenine nucleotides are in the sequence ATP > ADP > AMP (in other words, the greater the number of phosphato groups, the stronger the bonding) and also that the stabilities remain substantially unchanged when adenine is replaced by other purines.

The charge on the metal ion appears largely to determine the stability of complexes with ATP and this is indicative of 'metal-chain' complex forma-tion taking place via a coordinated water molecule. Fig. 7.2 shows two

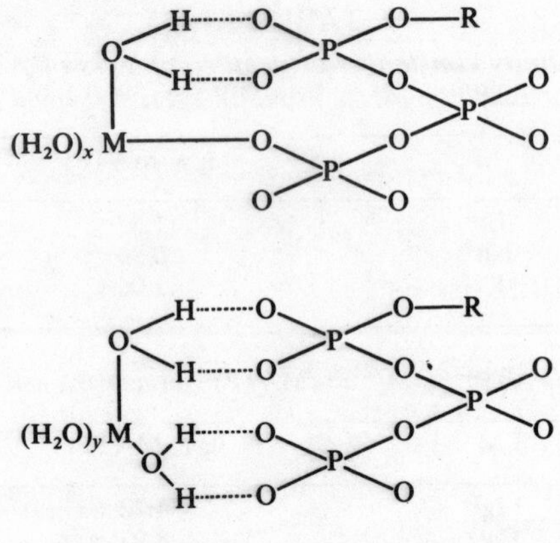

FIG. 7.2. Possible modes of linkage between a non-transition metal and an ATP^{4-} ion.

possible modes of linkage between a non-transition-metal ion and ATP^{4-} On the other hand, a metal ion like Zn^{2+}, with its known affinity for nitrogen donors, could be complexed as shown in Fig. 7.3, with the adenosine triphosphate ligand acting as a quadridentate chelate.

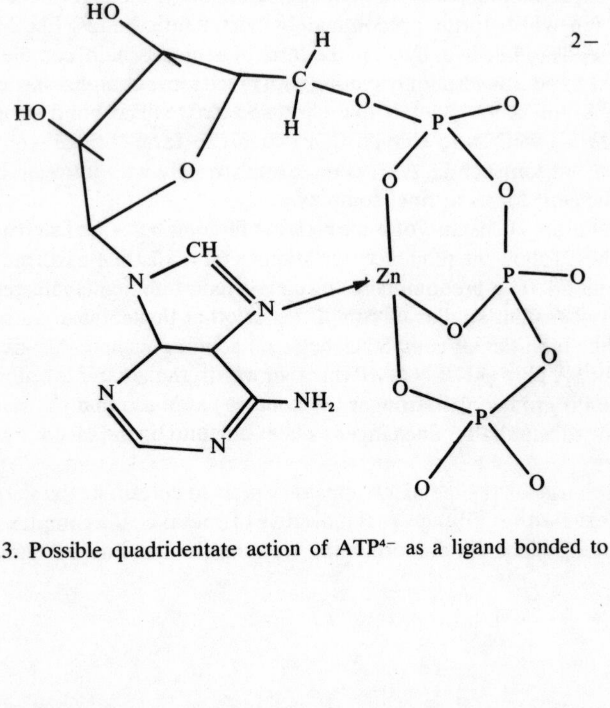

FIG. 7.3. Possible quadridentate action of ATP^{4-} as a ligand bonded to Zn^{2+}.

Metalloproteins

The 'trace' metals, manganese, cobalt, iron, copper, molybdenum, and zinc, are associated with protein structures in many molecules fulfilling important biological functions. In man, the most commonly occurring metal is iron (the average amount in a 70kg man is 4.1g); then come zinc (2.3g) and copper (0.2g). The other transition metals are present in amounts below 0.1g per 70kg.

All six trace metals are components of enzymes, protein substances which catalyse specific biochemical reactions. For example, zinc is found in carboxypeptidase, an enzyme which catalyses hydrolysis of peptide linkages. Iron, cobalt, copper, and molybdenum are essential constituents of certain enzymes in which catalytic activity is closely related to changes in the oxidation state of the transition metal. Thus iron is present in ferredoxin, an enzyme found in, for example, the chloroplasts of green plants, which acts as the initial electron acceptor of the photoactivated chlorophyll molecule. Metal ions are also present in coenzymes, essential materials for enzymatic activity. The cobalt complex, vitamin B_{12}, is an example of a coenzyme. Other metalloproteins perform functions of transport or storage of essential chemicals, for example, the oxygen is carried by the iron-protein complex, haemoglobin, and by copper-containing haemocyanin, and iron is stored by the storage protein, ferritin. In fact, the transition metals in living organisms are almost invariably associated with proteins. Examples given below will illustrate the importance of chelation processes which bind the metals firmly to the complex organic molecules.

A living cell requires various enzymes for the performance of particular functions. These species are of necessity highly specific in their catalytic behaviour, otherwise complications would ensue because of interference with other processes occurring simultaneously. Specificity is therefore very much higher than that usually associated with catalysts. It arises from the very special and complex structural relationships which exist between an enzyme and the substrate it activates.

In all cases where X-ray studies have been performed, it has been established that the enzyme possesses a cleft at the active site into which the substrate must fit to become activated. A particular configuration of the protein molecule is associated with the formation of such a cleft and it is this which is responsible for the specificity of the enzyme.

The strength with which the metal ion is bound by the protein structure varies from one enzyme to another. In some, the metal ion is strongly held and can be removed only under extreme chemical conditions; in others, the metal can be easily removed and replaced reversibly. In the latter case, different metals may be taken up by the enzyme structure but if this occurs the activity is altered, generally being lessened and sometimes disappearing altogether.

Carboxypeptidase A

Some important enzymes involved in hydrolytic reactions contain zinc. These include carboxypeptidase A and carbonic anhydrase.

The structure of carboxypeptidase A was the first to be determined for a metalloenzyme. Carboxypeptidase A is secreted in the pancreas as an inactive zymogen, procarboxypeptidase A (PCA), and is distributed in many species, ranging from dogfish to man. Four forms of carboxypeptidase A arise from enzymatic release from bovine procarboxypeptidase A, and detailed crystallographic work has been arried out on one of these, CPAα.

CPAα has a molecular weight of 34 472 and contains 307 amino-acid residues and one zinc atom. The enzyme specifically catalyses the hydrolysis of the carboxy-terminal peptide bonds in protein and peptide substrates. The peptide bond which is hydrolysed must be adjacent to a C-terminal free carboxylate ion and the rate of hydrolysis is enhanced if the side-chain of the C-terminal residue (R_1') is aromatic or branched aliphatic of L-configuration. Hydrolysis occurs at the position indicated by the dotted line.

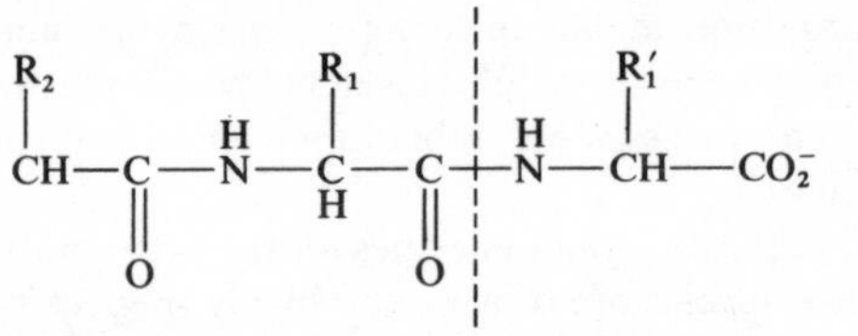

$$R_2 \quad\quad\quad R_1 \quad\quad\quad R_1'$$
$$\text{CH}-\text{C}-\overset{H}{\underset{}{N}}-\overset{}{\underset{H}{C}}-\text{C}+\overset{H}{\underset{}{N}}-\text{CH}-\text{CO}_2^-$$
$$\quad\quad \underset{O}{\|} \quad\quad\quad \underset{O}{\|}$$

Carboxypeptidase A also shows esterase activity, catalysing the hydrolysis of ester groups.

In CPAα, the zinc ion is coordinated by three amino-acid groups, His-69 and His-196 (via a nitrogen atom of the imidazole group of histidine) and Glu-72 (via the negatively charged oxygen of the glutamic acid group). Four-coordination in a very distorted tetrahedral configuration is completed by a water molecule.

When the enzyme binds to a substrate a number of interactions develop and there are major conformational changes in the structure of the enzyme molecule. Chief effects of the binding of a substrate by the active site in CPAα are:

(*i*)   displacement of the water molecule coordinated to zinc by the oxygen of the carbonyl group of the susceptible peptide link on the substrate;

(*ii*)   formation of a bond between the negatively charged terminal carboxylate group of the substrate with the positively charged guanidinium group of Arg-145;

(*iii*) insertion of the COOH-terminal side-chain into the 'pocket' of CPAα, thereby displacing several of the water molecules originally there: this region now becomes largely hydrophobic in character;

(*iv*)   movement of the Tyr-248 residue of CPAα by a twisting of the carbon-carbon chain so that its OH group is now only 300 pm from the susceptible peptide link of the substrate: probably this residue acts as proton donor to the NH group beyond nydrolysis;

(*v*)   movement of Glu-270: this is implicated in the hydrolysis reaction and may provide the base for the attack of the carbon atom of the susceptible peptide bond.

The environments of zinc and coordinated substrate are illustrated schematically in Fig. 7.4 in the case of glycyl-L-tyrosine.

One particular property of the zinc ion which makes it very effective in promoting hydrolysis is its strong polarizing power. Coordination of such an ion to the carbonyl oxygen of the substrate promotes hydrolytic fission

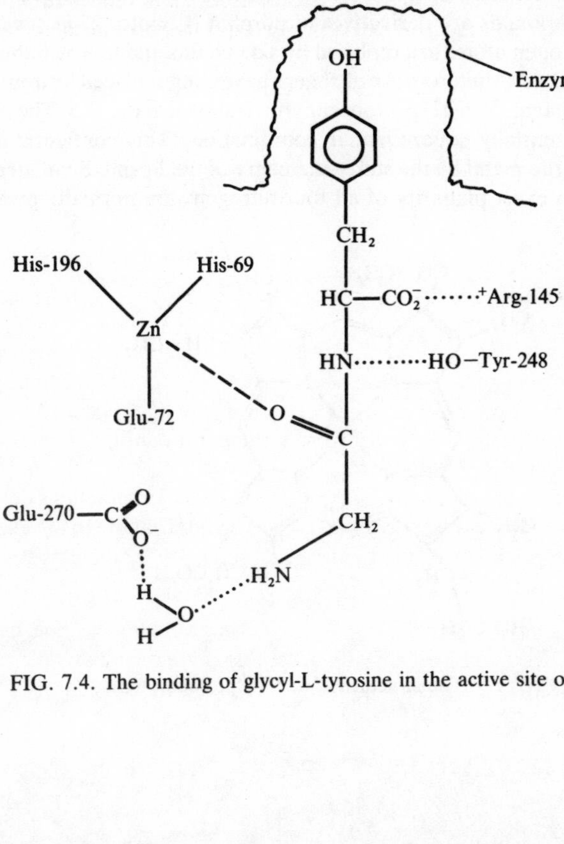

FIG. 7.4. The binding of glycyl-L-tyrosine in the active site of CPAα.

of the peptide link. Replacement of zinc by another metal affects the peptidase and esterase activities. The only metal to increase activity is cobalt(II); others either reduce it or render the enzyme completely inactive. The several conformational changes taking place in the protein chain of the enzyme when it interacts with a substrate impose critical stereochemical limitations on the site of the metal ion and the effectiveness of cobalt in promoting enzymatic activity is probably associated with its preference, like zinc, for 4-coordinate tetrahedral stereochemistry.

Another important zinc-containing enzyme is carbonic anhydrase, responsible for the hydrolysis of carbon dioxide; it is considered in detail in Phipps: *Metals and metabolism* (OCS 26).

Haem iron proteins

These proteins are associated with a variety of biological functions. They include haemoglobin, the oxygen-transport protein which gives the red colour to human blood; myoglobin, for oxygen storage; the cyto-chromes, involved in electron-transfer reactions; peroxidases, which catalyse oxidations by hydrogen peroxide; and catalases, which catalyse the disproportionation of hydrogen peroxide to water and oxygen.

These compounds are derivatives of porphin (Chapter 2) in which all pyrrole hydrogen atoms are replaced by side-chains and in which the two protons attached to heterocyclic nitrogens have been replaced by iron. The structure of haem, iron(II) protoporphyrin, is shown in Fig. 7.5. The metal ion is in essentially square-planar coordination. This configuration is forced upon the metal by the stereochemistry of the ligand. Small depart-ures from an exact planarity of all four nitrogens are normally present;

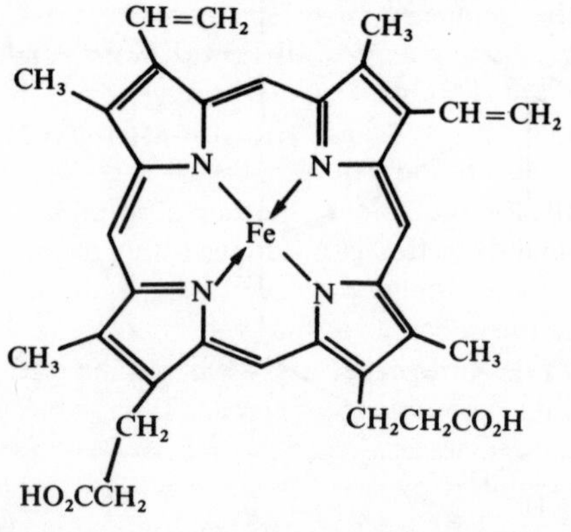

FIG. 7.5. The structure of iron(II) protoporphyrin.

these may become more significant when bulky substituents are present. With metals like iron, an extra one or two ligands can be coordinated in axial positions perpendicular to the plane of the porphyrin ring, thus forming a square-based pyramidal or a distorted octahedral complex.

Haemoglobin, myoglobin, the cytochromes, the peroxidases, and the catalases all contain the iron protoporphyrin group but they differ in the nature of the ligands in the axial positions. The specific biological reactivities of these metal complexes depend chiefly on the proteins with which the metal atom is linked in one or both of these positions. The sixth coordination position may be occupied by a ligand atom from a protein molecule or by a water molecule. In the latter case, the water molecule is usually readily replaceable by other small ligands like oxygen or carbon monoxide, as in haemoglobin or myoglobin.

The iron(II) protoporphyrins and iron(II) haem proteins are usually readily autoxidized to the corresponding iron(III) complexes. The ease with which oxidation is effected is dependent on the substituents present and many closely related iron haems are known which show slightly different redox behaviour. The redox potential of porphyrin complexes is very sensitive to the influence of substituents anywhere in the pyrrole ring system. The reason for this is the highly conjugated system linking together the four donor nitrogens. The extensive conjugation ensures that the electron-attractive or electron-dative properties of a substituent are transmitted to all four coordinating nitrogens and thence to the iron atom. Such effects can also be transmitted to ligands coordinated in the axial positions.

Conjugation also results in the characteristic electronic spectra of porphyrins. The porphyrins themselves and their metal complexes are intensely coloured, primarily because of transitions which occur at energies within the frequency range of the visible spectrum. The conjugated electronic system always gives rise to a very intense absorption band at about 400 nm. This band is known as the 'Soret' band and molar absorbances as high as 5×10^5 are observed for it in some complexes. Other less intense absorption bands occur at lower frequencies and are largely responsible for the colour. Further absorption bands are also found, arising from d-d electronic transitions within the metal ion or from charge-transfer transitions between the porphyrin molecule and the metal.

In haemoglobin, the iron atom of the haem structure is further coordinated by an imidazole nitrogen of a protein histidine group. The sixth coordination position is occupied by water and this can be replaced reversibly by other small molecules. When oxygen is coordinated in place of water, oxyhaemoglobin is formed. This is a low-spin iron(II) complex, with the iron atom situated in the plane of the porphyrin ring. The Fe–N bond length is 200 pm. In deoxyhaemoglobin, containing coordinated water instead of oxygen, the iron is in a high-spin state and its stereo-chemistry is significantly different. The high-spin iron(II) ion is too large to remain coordinated in the plane of the ring and it now lies above the plane of the nitrogen donors along the axis of coordination of the imidazole ligand. The Fe−N bond length increases to 290 pm. These structural changes are associated with other conformational changes within the haemoprotein structure.

Haemoglobin is actually a tetrameric molecule, containing four units, each consisting of a haem group with its coordinated protein. There are subtle stereochemical interrelationships between these units which, together with the structural changes within each haemoprotein unit that accompany oxygen uptake, provide a whole complex of interactions which are responsible for the physiological functions of this protein.

Vitamin B₁₂ and related compounds

Vitamin B₁₂ is a cobalt(III) complex which was first isolated in 1948. Subsequently, its crystal structure was determined by X-ray diffraction and its chemical properties intensively studied. Like other vitamins, it is required in small quantities by humans and other animals, to·perform specific functions. It is the only metal-containing vitamin and cannot be synthesized by the organism but must be obtained from the diet. In man, a deficiency of vitamin B₁₂ is associated medically with pernicious anaemia, a disease caused by the failure of the stomach to secrete hydro-chloric acid and the protein known as intrinsic factor, resulting in a failure to absorb vitamin B₁₂. A deficiency in vitamin B₁₂ affects the production of red and white blood cells in the bone marrow and this leads to anaemia.

Vitamin B₁₂ was first isolated from the liver (it had been known for a long time before this vitamin was discovered that sufferers from pernicious anaemia showed marked improvement when fed on a diet of raw liver) and now is produced commercially by fermentation processes using micro-organisms such as species of the genera *Streptomyces* and *Propionibacter* and also *Bacillus megatherium* and *Nocardia rugosa*. The most important large-scale use is as an additive to feedstuffs for pigs and poultry.

The molecular structure of vitamin B₁₂ is shown in Fig. 7.6. The central feature is the cobalt(III) ion, approximately octahedrally coordinated by the four nitrogens of the corrinoid ring, by a nitrogen of the benzimidazole residue of the nucleotide group, and by a cyanide ion. The corrinoid ring is non-planar; although conjugation in the corrinoid system is extensive it cannot continue completely around the ring because of the single C−C bond between two of the pyrrole rings. There is therefore no electronic require-ment for planarity and, as a result of the operation of intermolecular forces, the most stable structure is a buckled arrangement of the four donor nitrogens. Substitution of axial ligands in the corrinoids occurs

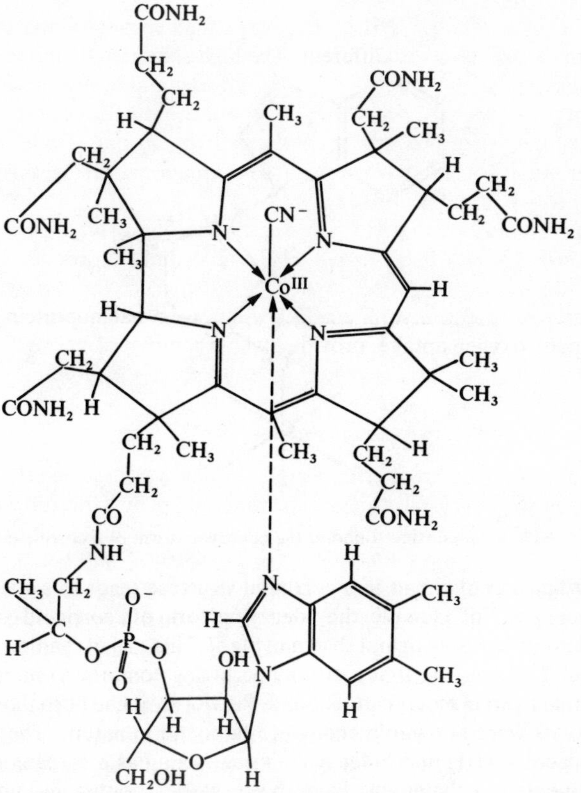

FIG. 7.6. The structure of vitamin B$_{12}$.

easily and, as a result, many related compounds can be synthesized. For example, a range of complexes is known in which the benzimidazole ligand is replaced by a water molecule. These are called cobinamides. The presence of cyanide ion in vitamin B$_{12}$ is an artefact introduced during the isolation of the complex from nature sources. Vitamin B$_{12}$ is known as cyanocobalamin and substitution of cyanide by other groups produces different cobalamins.

The diamagnetism of vitamin B$_{12}$ shows cobalt is in an oxidation state of +3. The complex is uncharged in neutral solution, the triple charge on the metal ion being counterbalanced by the negative charges on the cyanide ion, the corrin ring (which carries a single negative charge because the parent corrin structure loses one proton upon coordination), and the phosphate group of the nucleotide side-chain.

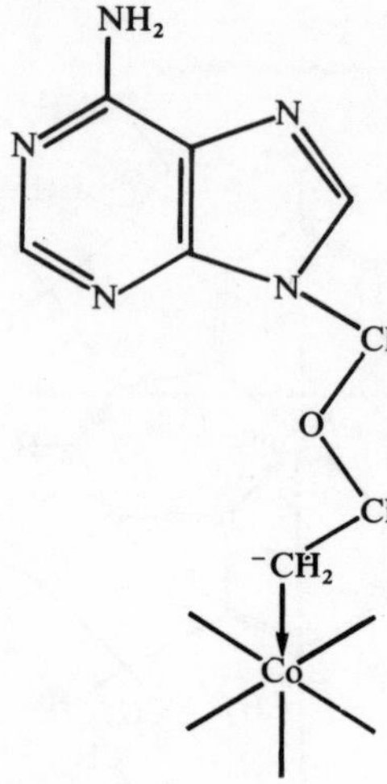

FIG. 7.7. 5-Deoxyadenosine found in the coenzyme form of a corrinoid structure.

Coordination of cobalt in a corrinoid structure leads to some unusual consequences. For example, the 'coenzyme' form of a corrinoid containing the 5-deoxyadenosine ligand shown in Fig. 7.7 instead of cyanide has been isolated. This was the first naturally occurring complex to be identified containing a direct metal-carbon bond. Previously it had been thought that such bonds were not stable enough to be found in nature. The complex contains cobalt(III) and 5-deoxyadenosine acting as a carbanion. Other organometallic cobalamins have been made, containing unidentate organo-ligands like CH_3CH_2-, $CH_2=CH-$, $HC\equiv C-$ and CH_3CO-. These organocobalamins have provided a valuable means for studying the effect of σ-bonded organo-ligands on the reactivity of other groups coordinate to cobalt, in contrast to the more commonly studied π-bonded ligands such as olefins and cyclopentadienyl.

Although most cobalt(III) complexes are octahedral, it has been established that organocobalt(III) corrinoid complexes can also adopt a square-pyramidal 5-coordination around the metal. This provides another example of how the nature of the ligand effects the stereochemistry adopted by the metal ion.

Finally, it is interesting to note that some quite simple analogues of the cobalt corrinoid complexes are to be found in the cobaloximes, complexes of dimethylglyoxime and cobalt. Bis(dimethylglyoximato)cobalt(III) contains square-planar cobalt coordinated by four nitrogens; further ligands are easily bonded in the axial positions. The similarity with the stereochemistry of cobalt in vitamin B_{12} is illustrated by the structure of

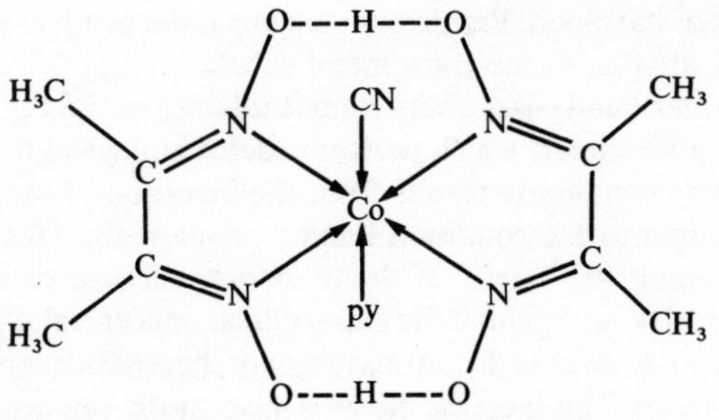

FIG. 7.8. The structure of cyanopyridine cobaloxime.

cyanopyridine cobaloxime (Fig. 7.8). The pattern of reactivity is broadly similar, for example, stable organo-derivatives can be readily prepared, and, as in the corrinoids, cobalt(III) can be readily reduced to lower oxidation states. As a consequence, the cobaloximes have proved to be useful model compounds whose reactions have given insight into the chemical behaviour of more complex molecules, such as vitamin B_{12}.

Chelating agents in medicine

We have already seen that there are many strongly chelating compounds present in biological material and that these compounds are characterized by the existence of highly specialized bonding sites, at least for iron and the other essential transition metals. In a living system, the concentrations of metal ions and their complexes are controlled within narrow limits. If the 'natural' balance of these concentrations is disturbed by internal or external causes then the organism can no longer behave normally and disorders result.

Thus a number of major diseases are associated directly with changes in the concentration of trace metal ions in certain tissues and body fluids. For example, in patients suffering from Wilson's disease, a disorder of the liver which affects the central nervous system, the natural mechanisms for the control of copper concentrations are disturbed. The amount of copper in tissues such as those of the liver, brain, and kidneys is observed to be much greater, in some patients as much as one hundred times greater, than average. Copper is stored in the liver in the form of copper proteins and it appears that the accumulation of free copper ion is the result of a deficiency of the copper protein ceruloplasmin.

A deficiency of any essential metal ion similarly disrupts the functioning of a living system. For instance, zinc deficiency results in reduced enzyme activity and the breakdown of normal metabolic processes. Probably the best-known example of the effect of metal deficiency is that of anaemia, which results from lack of iron and hence of insufficient haemoglobin for the red cells of the blood. Pernicious anaemia is the result of a deficiency of vitamin B_{12}, that is, of the trace metal cobalt.

On the other hand, organisms cannot tolerate unusually high concentrations of those metals which perform useful biological functions. Thus iron poisoning commonly results from the ingestion of excess iron compounds, leading to the condition known as siderosis. This is found, for example, among the Bantu of South Africa because of their habit of cooking in iron pots. Again, if the extracellular concentration of potassium ion is increased to double the normal amount, heart disorders and possibly even death result. This increase disturbs the osmotic balance of electrolyte between the exterior and interior of the cell.

The presence of some metals, such as lead, arsenic, and mecury, which apparently have no beneficial biological action, causes well-known toxic effects. In these cases, poisoning results in a big increase in the amount of copper and zinc excreted. It appears, therefore, that sufficient foreign metal ion is absorbed by the organism to displace the catalytically active metals from their binding sites on protein molecules and hence to disrupt normal enzyme action.

Some of the problems of metal toxicity may be overcome by use of *chelation therapy*. For this, chelating agents are used to remove toxic metals like lead, arsenic, or mercury or to reduce metals such as iron or copper to normal concentrations. The major problem is to achieve these ends without simultaneously reducing the concentration of other essential metal ions below the limits of safety. Ideally, chelation therapy results in the removal, in the form of a metal chelate, of much of the toxic metal present in the blood-stream. The subsequent removal from other parts of the body will depend primarily on the rate at which the metal becomes redistributed within the tissues.

For use in chelation therapy, a chelating agent should be of low toxicity, not easily metabolized and capable of penetrating to metal storage sites. It should also preferably be capable of coordinating with a metal ion to form several chelate rings, that is, to give a highly stable complex, and to be fairly selective in its chelation.

Lead poisoning

Lead is a cumulative poison in man and it affects the central nervous system: it is particularly liable to cause brain damage in children. Lead poisoning is a hazard associated with a number of industrial processes. In parts of Britain where lead piping is still in use for carrying domestic water supplies, lead poisoning can also result from the uptake of lead from the drinking water. For the relief of symptoms of lead poisoning, the most useful chelating agents have been found to be EDTA and other aminopolycarboxylic acids.

The monocalcium disodium salt of EDTA is used in the standard procedure for the treatment of lead poisoning. This form is used rather than one of the sodium salts of EDTA because the addition of calcium with the chelating agent avoids problems of calcium-ion depletion. This would otherwise ensue because of the tendency to form the stable water-soluble calcium-EDTA complex.

In acute cases of lead poisoning, intravenous injection of $CaNa_2$-EDTA leads to a rapid excretion of the lead-EDTA complex into the urine: EDTA is not readily metabolized and most of the injected calcium complex is, therefore, excreted in the urine. The non-selectivity of this kind of chelating agent is illustrated by the fact that this treatment also leads to an increased excretion of copper and manganese, together with the depletion of zinc reserves in the body. It is then necessary to ensure that, after chelation therapy, the concentrations of these essential ions are restored to normal by the administration of suitable preparations containing them.

Arsenical poisoning

Arsenic is another element that has characteristic toxic effects on human organisms. Arsenic is known to disturb the pyruvate oxidase system and the mode of action is believed to involve chelation of arsenic(III) with suitably arranged thiol groups on the enzyme:

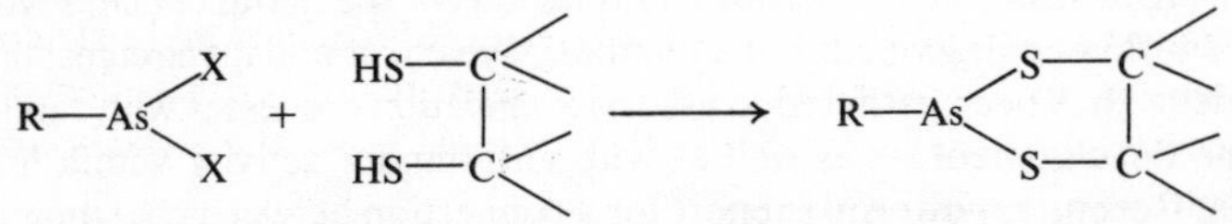

The inhibition of pyruvate oxidase in this manner leads to characteristic biochemical lesions, for example, the formation of skin blisters which is one of the primary effects of the poison-gas Lewisite, $Cl-CH=CH-AsCl_2$.

The tendency to react with sulphur donors indicates the 'soft' nature of arsenic; it shows no comparable tendency to complex with nitrogen donors. Recognition of this preference for bonding to sulphur led to the development of the chelating ligand 2,3-dimercaptopropanol (British Anti-Lewisite — BAL) as an antidote to Lewisite. This reacts with arsenic to form a five-membered chelate ring and effectively prevents its interaction with the essential thiol groups of enzymes such as pyruvate oxidase.

BAL and other similar dithiols are also effective in the treatment of poisoning by heavy metals such as mercury which, like arsenic, show a preference for bonding to sulphur. Some disadvantages are associated with the use of BAL, particularly its unpleasant smell, ease of oxidation, and instability in aqueous solution. However it does not complex alkaline-earth metal cations to a significant extent and, unless chelated, it decomposes rapidly in the body so that its use does not result in the removal of essential metal ions.

For the removal of toxic excess of copper, arising, for instance, in Wilson's disease, the D-isomer of penicillamine is the most effective chelating agent. This is able to coordinate to a metal via sulphur, nitrogen, and oxygen donors. It is water-soluble and stable and appears to be able to remove excess copper in the form of soluble complexes without depletion of the normal stores of copper ion within the body.

$$CH_2SH$$
$$|$$
$$CHSH$$
$$|$$
$$CH_2OH$$

BAL

$$CH_3$$
$$|$$
$$HS—C—CH_3$$
$$|$$
$$H_2N—C—CO_2H$$
$$|$$
$$H$$

D-penicillamine

Chelating agents in the treatment of cancer

Trace elements essential for many biological processes are, with few exceptions, transition metals. These are pre-eminently the metals that form highly stable chelates, and the importance of chelation in biology can be gauged from the properties and functions of the metal chelates which have just been described. For a number of years, abundant circumstantial evidence has been gathered to associate chelation processes with carcinogenic development — as well as with anti-tumour activity within living cells. Recently, powerful support for a connection between chelation and cancer has come from the discovery that some of the coordination complexes of platinum are very effective inhibitors for the growth of tumours.

Most chemotherapeutic agents effective against cancer are actually chelating agents. For example, 2,2'-bipyridyl and 1,10-phenanthroline are known to possess anti-tumour activity. A compound which shows anti-tumour activity in one environment generally is able to function as a carcinogen in another. It follows that most chemicals which have been implicated in the development of cancer are also chelating agents. These include azo dyes such as *p*-dimethylaminoazobenzene or their metabolic products, which probably act as carcinogens by chelation mechanisms.

p-dimethylaminoazobenzene

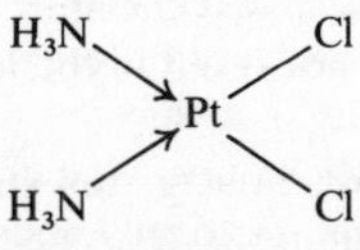

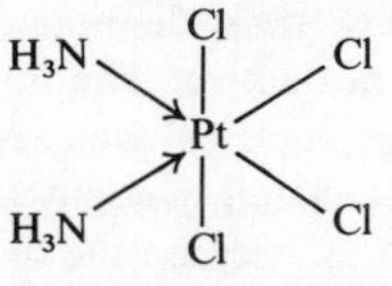

cis-diamminodichloroplatinum (II)   cis-diamminotetrachloroplatinum (IV)

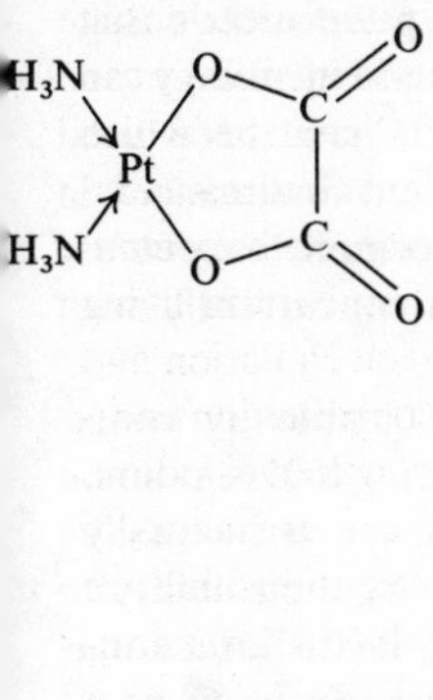

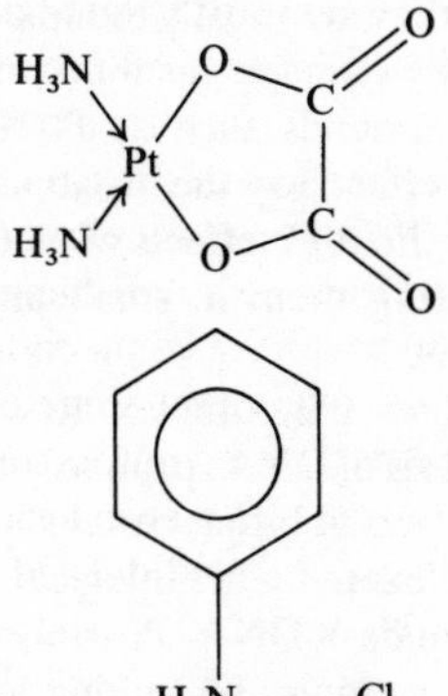

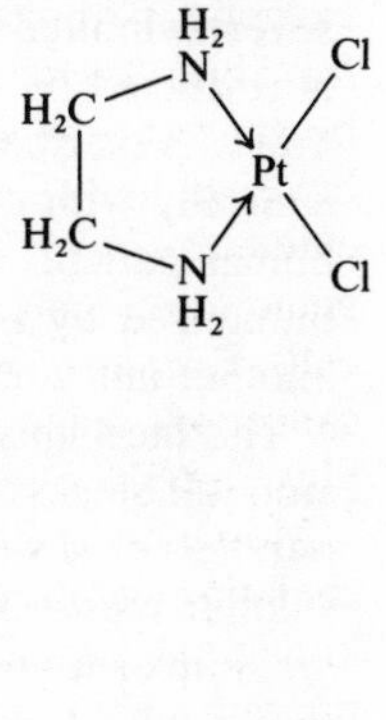

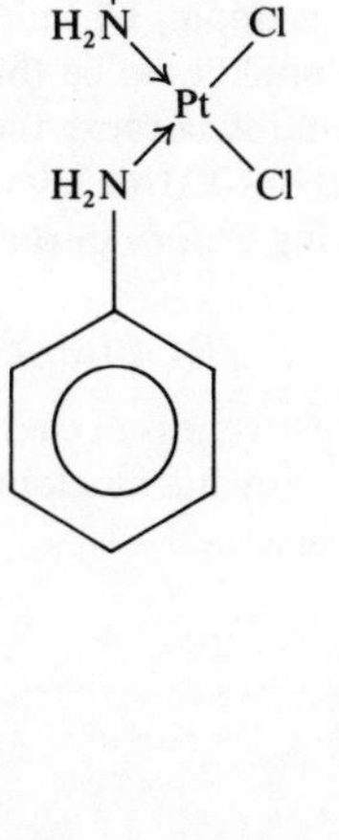

In 1964, the observation was made that, when an alternating electric field was applied using platinum electrodes across a cell in which the bacteria *Escherichia coli* were growing, all cell division was inhibited. The production of DNA was not inhibited and growth continued longitudinally with the formation of long filaments. The effective agent for blocking cell division was identified as some platinum-containing complexes in solution at low concentrations (about 10 p.p.m.). These, produced electrolytically by the applied field in the cell, were identified as *cis*-diamminodichloroplatinum(II) and *cis*-diamminotetrachloroplatinum(IV). They are both active in the inhibition of tumour growth. Other platinum(II) complexes which show anti-tumour activity include the four shown below. The striking feature of all these complexes is the *cis*-arrangement of identical ligands. This stereochemistry appears to be essential for anti-tumour activity. Thus *trans*-diamminodichloroplatinum(II) and other *trans*-isomers are not active. The implication of the very different behaviour of the two geometrical isomers of Pt(NH₃)₂Cl₂ is that the anti-tumour activity is associated in some way with reactions involving chelation. In *cis*-Pt(NH₃)₂Cl₂, because the two chlorines are less firmly bound than the two ammonia ligands, their replacement by a chelating agent is possible. Replacement of both chlorines in the *trans*-isomer by a chelating ligand is stereochemically most unlikely because the great majority of chelating agents cannot effectively coordinate *trans*-positions in the coordination shell of a metal.

Whatever the nature of the inhibiting reaction, there appear to be very subtle requirements concerning the properties of those platinum complexes which are effective. For instance, the complex has to be an uncharged species; no activity has been associated with any charged *cis*-complex. Again each *cis*-complex has to be tested on an individual basis from the point of view of its cancer-reducing capability. For instance, [Pt(NH₃)₂(Ox)] and [Pt(NH₃)₂(en)] are effective anti-tumour agents but [Pt(Ox)(en)] is highly toxic and cannot be used. In contrast, both [Pt(NH₃)₂(mal)] and [Pt(en)(mal)] have anti-tumour properties, and the latter is much less toxic than its oxalato analogue.

The number of *cis* platinum complexes which can be used in practice is severely limited because they are mostly too toxic. The permissible dosage is restricted by the kidney damage sustained if the amount used is too great. Nevertheless, compounds such as Pt(NH₃)₂Cl₂ have been used clinically with dramatic effect for the treatment of terminal cancer in human patients. Possibly the toxic effects of platinum complexes could be minimized by administering them in small amounts concurrently with another anti-tumour agent.

The mechanism by which the complex acts is not completely understood. It blocks the synthesis of DNA, primarily by binding to DNA. Some parallels exist with the action of bifunctional alkylating agents such as the sulphur mustards. These exert their biological action by their ability to cross-link sites on two strands of DNA. Possibly *cis*-Pt(NH₃)₂Cl₂ functions in a similar fashion, for example, by linking amino groups on adjacent DNA strands. These are unlikely to be the same sites as those linked by the action of sulphur mustards because the replaceable chloro ligands in *cis*-Pt(NH₃)₂Cl₂ are only about 300 pm apart compared with the distance of 800 pm between the leaving groups in the sulphur mustards.

PROBLEMS

7.1. In a study of the effectiveness of chelating agents in removing heavy metals complexed with haemoglobin, the following results were obtained:

Chelating agent	Percentage mercury removed	Percentage zinc removed
Penicillamine	76.3	43.9
N-Acetylpenicillamine	80.4	32.0
EDTA	0.5	92.9
DTPA	3.1	90.2

In each case, the molar ratio of metal ion:haemoglobin is 1:1. Comment on these figures in relation to the coordinating properties of the above ligands with mercury and zinc ions. What do they suggest about the mode of bonding of mercury in its complex with haemoglobin?

7.2. Dimethylglyoxime ($C_4H_8N_2O_2$, 1.5 mole) and cobalt(II) chloride hexahydrate (0.5 mole) were mixed together in methanol, then sodium hydroxide (1.5 mole) and pyridine (0.75 mole) were added, the suspension was cooled to 268 K, and further sodium hydroxide and some sodium tetrahydroborate (0.13 mole) added. Finally, dimethylsulphate (0.8 mole) was added. The solution after evaporation of the solvent yielded an orange crystalline solid A. Percentage analysis of A: C = 43.86; H = 5.84; N = 18.67.

When A was warmed with dimethylsulphoxide in aqueous methanol, orange crystals of another complex, B, were obtained. Percentage analysis of B: C = 33.51; H = 6.03; N = 17.48.

B was decomposed by the action of warm concentrated potassium hydroxide solution, methane being among the decomposition products. When B was heated, water was driven off.

What are the formulae of A and B? Draw diagrams to represent their structures.

8. Industrial applications

Sequestration of metal ions

In many industrial processes the presence of adventitious metal ions causes problems because of the adverse effects these can have on product quality. Such difficulties are avoided if the concentration of the free metal ion is reduced to such a level that at least some of its chemical reactions are no longer observed. An effective means of achieving this is by the addition of a suitable chelating agent which, by complexing with the metal ion, masks its characteristic reactions. The process by which a soluble complex of a metal is produced so that, for example, it is no longer precipitated by anions which form sparingly soluble salts with the free metal ion, is known as sequestration (derived from the Latin *sequestrare*, to commit for safe keeping) and the complexing species which bring this about are called sequestering agents.

Most of the raw materials used by industry contain metal ions. For example natural water, used so extensively for washing operations and as a medium for chemical reactions, contains dissolved metal salts, especially those of calcium and magnesium, which are chief contributors to the hardness of the water. If the water has been in contact with soil it will normally also contain certain polysaccharides and their breakdown products such as humic and fulvic acids. These organic polymeric molecules have chelating properties and are probably complexed to some extent with naturally occurring metal ions. Therefore they serve as a means of introducing such ions into an industrial process.

Raw materials originating in plants always contain those metals, such as magnesium and manganese, that are essential for growth. Most naturally occurring organic materials used industrially are, by their constitution, good chelating agents, and provide sources of metal ions. For example, wool is a complex protein containing coordinating groups such as $-OH$, $-NH_2$, $-CO-NH-$, $-Co_2H$, $-SH$, and $-S-S-$. It has a capacity for binding metal ions comparable with that of some synthetic ion-exchange resins.

Mineral substances used industrially are also normally contaminated with metallic salt impurities. For example, the lime commonly employed for neutralization purposes contains at least traces of some other metal ions besides calcium.

Corrosion of industrial equipment is a common source of contamination by iron and, to lesser extents, by metals such as zinc, tin, aluminium, lead, copper, nickel, and cobalt.

Unless steps are taken to sequester metal ions introduced from any of the above sources, a number of deleterious consequences can follow. These include the formation of slightly soluble metal salts which could lead to difficulties, for example, in textile-processing operations like bleaching and dyeing, or result in the deposition and build-up of scale in boilers; the catalysis of unwanted side-reactions such as, for instance, the oxidation of fats or oils to produce rancidity, poor colour stability, and general deterioration; or reaction with chemicals subsequently introduced, as in the case of many organic dyestuffs used in the textile industry which themselves behave as chelating agents and can react with metal ion impurities to give undesired coloured products or precipitates.

Although in principle all types of chelating compounds could be employed for the purpose of sequestration, in practice only three kinds are used extensively in industry. These are the polyphosphates, the amino-polycarboxylic acids, and some hydroxycarboxylic acids. In almost all applications, sequestration of all metals present is desirable and so there is no requirement for a chelating agent which acts selectively.

Polyphosphates

These are the most widely used sequestering agents in industry. They were the first to be used for water softening and for large-scale usage are still often preferred, because of their comparatively low cost, to organic complexing agents.

Two types of polyphosphate, crystalline and glassy, are known. In some crystalline polyphosphates, chains of phosphate units exist, linked together by shared oxygens. Typical of these are sodium tripolyphosphate ($n = 1$) and sodium hexapolyphosphate ($n = 4$). In others, cyclic polymeric anions exist, as in sodium trimetaphosphate. The glassy polyphosphates contain much longer chains.

All polyphosphates have the ability to form water-soluble complexes with metal ions, like those of the alkaline earths. The metal is chelated by negatively charged oxygen atoms attached to adjacent phosphorus atoms. Further coordination with oxygens of the same chain or another polyphosphate anion also occurs.

sodium trimetaphosphate   polyphosphate chain

The sequestering power of a chain polyphosphate increases with increasing chain length. Thus a smaller weight of sodium hexapolyphosphate than of sodium tripolyphosphate is required to mask a certain quantity of calcium. This suggests greater multidentate character for the longer chain but the structural complexity of the larger polymer makes it difficult to decide how many chelate rings are formed in its calcium complex

The sequestering property of a polyphosphate depends on the pH. A greater quantity of any polyphosphate is needed to complex with a given amount of metal at high pHs (greater than 11). This is primarily because there is a great tendency for the formation of metal hydroxides in such alkaline solutions.

Polyphosphates are used extensively as additives in soaps and detergents to prevent the formation of insoluble calcium and magnesium soaps. If this occurs in the washing of fabrics, then the further processing of these (bleaching, dyeing, etc.) can lead to unsatisfactory products.

Some formulations of detergents contain between ˙35 per cent and 50 per cent of sodium tripolyphosphate. This acts as a detergent 'builder'. Although it possesses little or no detergency itself, a builder, in the presence of chemicals having active detergent properties, markedly improves their performance. Alternatively, its use allows the active components to be used at much smaller concentrations. This is primarily a result of the sequestering capacity of the builder. It also has the property of dispersing insoluble materials by the formation of colloidal suspensions.

The chief chemical disadvantage of polyphosphates is their tendency to undergo hydrolysis, or *reversion*, to orthophosphate. Reversion results in a loss of complexing power and, at the same time, produces an anion which forms insoluble salts with most metal ions. The rate at which reversion occurs depends on the pH, the temperature, and the degree of dehydration of the polyphosphate ion. The reaction is catalysed both by hydrogen ion and by metal ions. When polyphosphate solutions are stored or heated for long periods, reversion tends to occur. The shorter-chain polyphosphates, for example $Na_5P_3O_{18}$, tend to revert more slowly than the longer-chain ones, so the former are preferred in practice.

The use of synthetic detergents accounts for at least half the phosphate content found in waste waters. Two major environmental effects are now identified with the widespread use and disposal of phosphates. These are eutrophication of lakes and waterways, and the contribution these compounds make to the growth of algae. As a result, the phosphate content of detergents is now limited by law in some parts of the world and much attention has been paid to the development of other builders which do not adversely affect the environment.

Complexones

Salts of EDTA, particularly the tetrasodium salt Na_4Y, are used at different stages of many industrial processes for the sequestration of metal ions. Such compounds are very effective because of the great stability of their complexes with metals. Thus the complexes with calcium and magnesium ions have larger stability constants than those involving other generally available chelating agents like polyphosphates. Moreover, the ability to sequester alkaline-earth metal ions remains virtually constant over a wide pH range in contrast to that of the polyphosphates.

Other advantages possessed by EDTA salts are the stability of their aqueous solutions to prolonged heating and their chemical inertness towards alkali.

Complexones are more expensive to produce than the other sequestering agents used on a large scale, but their greater cost is partly offset by the smaller amounts required. In applications involving organic compounds, complexones also have the advantage over inorganic ligands of greater compafibility with the material being treated.

One common application of EDTA is for the prevention or removal of scale in boilers. When, over a period of several months, hard water containing dissolved calcium and magnesium salts is used in boilers, insoluble compounds such as carbonates, phosphates, silicates, and sulphates are deposited as scale. These can be removed by the addition of Na_4Y to the boiler feed-water to render the scale soluble. Alternatively, the use of EDTA from the start prevents the formation of scale.

Applications of EDTA in the detergent industry include the following:

(*i*)   in liquid soap products and shampoos to maintain clarity and enhance performance by preventing or reducing the formation of calcium soaps;

(*ii*)   in bar soaps to prevent rancidity or discoloration caused or catalysed by trace metals;

(*iii*) in synthetic detergents containing sodium perborate or persulphate to stabilize and control the performance of these agents;

(*iv*)   in wax-removing floor cleaners to help solubilize the wax and maintain solubility of the amine soaps used as wax emulsifiers;

(*v*)   for cleaning glass and metal containers.

The complexone is invariably a minor component of the detergent formulation, usually between 1 per cent and 5 per cent. EDTA has been the most commonly used sequestering agent of this type although the characteristics as detergent builders of other complexones such as NTA have also been exploited commercially.

NTA has the economic advantage over EDTA that, on a weight basis, less of it is needed to complex cations in a 1:1 mole ratio, and also that the raw materials from which it is manufactured are less expensive than those required for EDTA. On the other hand, the stability constants of metal/NTA complexes are significantly smaller than those of their EDTA counterparts, so in some conditions an excess of NTA over that stoichiometrically required may have to be used to effect adequate sequestration.

Although it was at first believed that there were no particular problems associated with the disposal of waste water containing complexones, further study has indicated the likelihood of harmful effects in some circumstances. Thus NTA is normally degraded in waste-treatment systems and in the environment, but it does not degrade in anaerobic conditions, for instance, in septic tanks. Furthermore, NTA bonded to metals is believed to be non-biodegradable. The danger lies in its complexing power towards toxic metals like lead and cadmium. When these are mobilized as water-soluble complexes, there is a greater likelihood of the uptake of these metals by plants. This could provide a route whereby toxic metals enter the food of animals and humans. The mobility of these complexes could lead, for example, to the transmission of such metals across the placental barrier to the foetus in pregnant women and hence to an increased chance of birth defects.

Hydroxycarboxylic acids

The most important hydroxycarboxylic acids used as industrial sequestering agents are gluconic and citric acids and, to lesser extents, tartaric and saccharic acids.

citrate chelated to a metal ion

tartrate chelated to a metal ion

In neutral or weakly alkaline solutions, the citrate ion is found to be the most effective. Its metal complexes are more stable than those of the other acid anions, partly because citrate carries a greater negative charge (-3) than tartrate and saccharate (-2) or gluconate (-1). Another reason is the stronger coordinating power of the carboxylate group compared with that of the alcoholic OH group. The coordination of citrate to a metal ion involves two carboxyl groups and one hydroxyl group, whereas coordination with tartrate occurs between one carboxyl group and two hydroxyl groups.

The gluconate ion contains only one carboxylate group and this is involved, together with one or two hydroxyl groups, in the chelation of a metal ion. The observed order of effectiveness in chelation, citrate > tartrate > gluconate, is therefore in accordance with expectation from the structures and charges of the ligand anions concerned.

The uses of citrate and other hydroxy acids as sequestering agents include the removal of rust and scale from iron, the cleaning of surfaces of other metals such as copper and aluminium, the stabilization of frozen fruits by preventing the metal-catalysed oxidation of ascorbic acid, and their inclusion in solutions for electroplating and metal-finishing processes.

Compared with complexes of aminopolycarboxylic acids, those of hydroxycarboxylic acids have low stabilities and so these compounds are relatively inefficient for sequestration, at least in solutions where the pH is below 11. Much of their industrial importance lies in the fact that when the pH is above 11, and in the presence of free caustic alkali, protons ionize from the hydroxyl groups to produce more negatively charged ions which have strong chelating properties. As a result, sodium gluconate and salts of the other hydroxy acids are very effective sequestering agents in solutions of high pH for ions such as calcium and, in such circumstances, are considerably better than the inorganic polyphosphates.

Iron is one of the most frequently-encountered metal contaminants in industrial processes. Iron(III) has a strong tendency to hydrolyse and is converted in alkaline media to sparingly soluble hydrated iron(III) oxide. In many situations, it is most important to maintain the iron in soluble form. For example, cotton cloth requires boiling in 3-5% caustic soda solution to remove various impurities which would otherwise adversely affect its tensile strength and dyeing properties. If iron is deposited during this process, the whiteness of the cloth is spoilt by staining, and the metal will catalyse the decomposition of peroxide bleaches to such an extent that oxidation of cellulose takes place in the region of the stain. This produces an oxycellulose which lowers the tensile strength of the fibres. As mentioned in Chapter 5, EDTA is not an effective chelating agent for iron where the pH is above 8. HEEDTA is more usful up to pH 11 or 12 but neither is effective in solutions of pH > 12 containing 1% or more sodium hydroxide. For these strongly alkaline media, the best chelating agents are hydroxycarboxylates such as sodium gluconate or saccharate. Alternatively, triethanolamine is useful for iron sequestration. The effectiveness of any of these chelating agents is, in each case, associated with strong complex formation between iron(III) ions and ionized alcoholic OH groups.

Extraction and separation of metals

Many traditional processes for the extraction and purification of metals from their ores are pyrometallurgical. They are most suited to the treatment of high-grade ores or ore concentrates. With the depletion of many high-grade ores, methods of ore concentration have become more significant and there is an increasing requirement to process low-grade materials. Pyrometallurgical operations can be quite uneconomic for these purposes. Additional costs are likely to be incurred in the case of pyrometallurgical processes that lead to the emission of atmospheric pollutants, for example, the formation of sulphur dioxide in the roasting of sulphide ores, because of the extra operations needed to prevent the pollution exceeding legally defined limits. There has accordingly been much development in the treatment of ores by hydrometallurgical and solvent extraction processes. These employ aqueous or organic media at essentially ambient temperatures and so are inherently inexpensive from the viewpoint of energy consumption, as well as being relatively non-polluting compared with pyrometallurgical extractions.

Solvent extraction using chelating agents is now applied on a large scale in the extractive metallurgy of copper. Low-grade sulphide ore containing the metal can be leached with either an acidic liquor (dilute sulphuric acid) or an ammoniacal ammonium carbonate solution. In the presence of atmospheric oxygen, copper is extracted as water-soluble salts. In the early 1960s, General Mills Chemicals Inc., in the United States, developed the use of liquid ion-exchangers for the extraction of copper from leach liquors. The first ion-exchange compound to be used commercially for this purpose was 5,8-diethyl-7-hydroxydodecane-6-oxime (LIX 63). This

LIX 63

liquid extractant carries functional groups on two adjacent carbon atoms, that is, suitably placed for chelation of a metal ion following the loss of a proton.

LIX 63 is a suitable extractant for the removal of copper from solutions of pH 4 and above and so can be used to treat the product from the ammoniacal leaching of copper ores. It is not applicable to the more commonly obtained acidic copper leach liquor which has pH 1-3. Another chelating ligand was therefore developed to fill this need. This, designated LIX 64, contains a phenolic hydroxyl group and an oxime group, which are

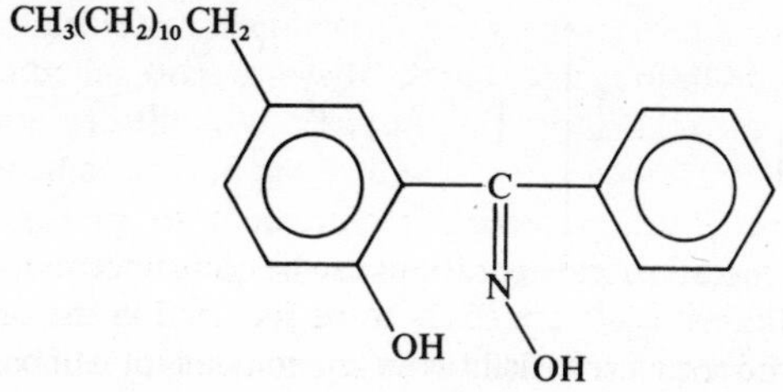

CH$_3$(CH$_2$)$_{10}$CH$_2$

LIX 64

able to chelate copper ion. The phenolic OH group has greater acidity than the alcoholic OH group in LIX 63, so complexation of copper with LIX 64 proceeds at a lower pH. Extraction of copper from leach liquor of pH $\simeq 2$ is quite feasible with this extractant. As the rate of extraction using LIX 64 is low, another compound, LIX 64N, containing a shorter hydrocarbon chain linked to the aromatic system is preferred in practice because this gives more rapid extraction.

The extraction proceeds according to the general equation:

$$2RH(org) + CuSO_4(aq) = R_2Cu(org) + H_2SO_4(aq)$$

After the extractant has been separated from the spent leach liquor, copper can be recovered into an aqueous phase by back-extraction with dilute sulphuric acid.

The selectivity of LIX 64N for copper is very high compared with that for other metals with which copper is commonly associated. Iron(III) is the only other metal extracted to a significant extent between pH 1.5 and pH 2.8 and, even so, the ratio of the amount of copper to that of iron extracted is of the order of several hundred to one.

LIX 64N is being used in several commercial plants in the United States for the extraction of copper from acid leach liquors. A very large plant is being built in Zambia to process about 180 000 kg of copper per day. Progress in the application of this technique is rapid and it is probable that, in a few years' time, much of the world production of copper will involve solvent extraction with chelating agents.

Boron is another element which is extracted commercially by use of a chelating agent. Aliphatic or aromatic diols in a solvent such as kerosene when contacted with an alkaline brine which contains borate will extract this in the form of a bis(diol)borate anion in association with an alkali-metal ion:

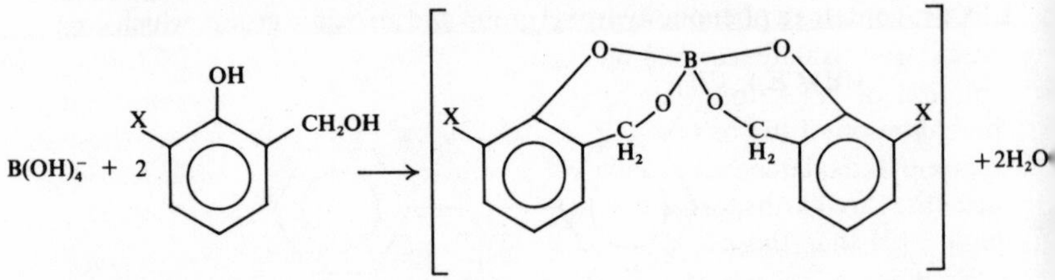

The borate can be recovered easily as an aqueous solution of boric acid and alkali-metal salt by contacting the loaded extractant with dilute mineral acid.

It is interesting to note that this type of chelation reaction forms the basis of a long-established volumetric procedure for the determination of boric acid. Boric acid itself is too weak for titration with sodium hydroxide solutions but in the presence of diols such as mannitol, it forms chelate complexes which are strong acids. The addition of mannitol to a boric acid solution thus makes it possible to carry out a direct titration with standard alkali using a conventional acid-base indicator.

Separation of the lanthanides

Because of similarities in their ionic size and chemical properties, several lanthanide elements are always found together in rare-earth minerals. As a consequence, the separation and isolation of individual lanthanides present considerable problems. A major motivation for devising an effective separational procedure has been to obtain macro quantities of the pure elements so that their physical and chemical properties can be thoroughly studied and applications for the individual metals can be developed. Chelating agents have been used in several ways to achieve separations of the rare earths in a very pure state.

Classical separational methods applied to this problem involved either fractional crystallization or fractional precipitation. For example, the separation of lanthanum, praseodymium, and neodymium can be achieved by fractional crystallization of the double nitrates with ammonium nitrate, $Ln(NO_3)_3.2NH_4NO_3.4H_2O$. Fractional precipitation requires the use of a deficiency of precipitating agent. An example is the separation effected between the 'light' rare earths (lanthanum to europium) and the 'heavy' rare earths (gadolinium to lutetium) and yttrium by fractional precipitation of their double sulphates with sodium sulphate. The double sulphates of the light rare earths are less soluble than those of the heavy elements and are precipitated first. The bulk of the heavy rare earths remains in solution if a deficit of sodium sulphate is added.

Chelating agents such as NTA and EDTA have been used to promote the effectiveness of fractional precipitation methods. Thus a mixture of rare-earths oxalates can be fractionated by dissolution in a weakly alkaline solution of NTA followed by careful acidification to cause fractional precipitation of the oxalates. In a solution containing a mixture of those rare-earth oxalates, which have approximately the same solubility products, the rare earths forming weaker complexes with NTA precipitate at a higher pH than those forming more stable complexes.

Solubility differences between salts of a given anion and a group of rare-earth ions are usually quite small. In order to achieve the separation of an individual element from all others present it may therefore be necessary to repeat many times, perhaps as many as several thousand, the fractional crystallization or precipitation operations. Even then, the isolated rare earth is likely to contain small amounts of one or more of the other lanthanides because of co-precipitation. To overcome this purity-limitation inherent in precipitation techniques, and to achieve more rapid separations, ion-exchange chromatography and counter-current solvent extraction techniques have been applied to the separation of the rare earths. Of these two, ion-exchange chromatography is the more efficient and this is now used industrially for the production of pure rare earths on a kilogram scale.

As a rule, cation-exchange resins show greater affinity for a metal ion the greater its positive charge and the smaller the size of the hydrated ion. Thus the tripositive ions of the rare earths are all strongly held by a resin, although the affinity decreases from La^{3+} to Lu^{3+}. In fact, the sizes of the hydrated ions are so similar that the resin shows little selectivity towards individual rare earths and the ion-exchange process itself is not very effective for achieving separations. However, if chelating anions are introduced into the aqueous phase, then the distribution of the metal ion between the solution and the resin phase is the resultant of two effects, its affinity for the resin and its affinity for the ligand anion, and experimental conditions favourable for separation can be defined.

One of the most widely used chelating agents for this purpose is ammonium citrate. The mixture of rare-earth ions is sorbed as a narrow band at the top of a cation-exchange resin column (in the hydrogen-ion form) and ammonium citrate solution is passed down the column. The individual rare earths move through the column as elution proceeds, the rate of movement being governed by the affinity of the ion for the resin and the stability of its complexes with citrate ion.

The rare-earth ion M^{3+} is distributed between the solution and resin phases in accordance with equilibria such as:

$$M(R) + (NH_4)_3Cit \rightleftharpoons MCit + (NH_4)_3(R)$$
$$MCit + Cit^{3-} \rightleftharpoons MCit_2^{3-}$$

(R represents the resin phase carrying negative charges).

As the citrate complexes come into contact with the resin, they tend to dissociate, hydrogen ion is displaced from the resin, and the metal ion is resorbed:

$$MCit + H_3(R) \rightleftharpoons M(R) + H_3Cit.$$

The rare earths resorbed first are those which form the weakest citrate complexes. The displacement of hydrogen ion lowers the pH of the solution and increases the tendency for the more stable citrate complexes to be broken down. The cations released are sorbed lower down the column. The heaviest lanthanide, which forms the strongest citrate complex, moves furthest down the column before being taken up again by the resin. As elution is continued with fresh ammonium nitrate solution, the pH of the solution in the column slowly rises until the metals are desorbed again and the above cycle is repeated.

Separation of fission-product rare earths at tracer levels and of mixtures of rare earths for analysis can be successfully achieved using citrate as complexing gent.

Separations on a larger scale are not straightforward if citrate is used as eluent. It cannot be used at concentrations above 5×10^{-3} mol dm^{-3} at pH 8.0, otherwise the slightly soluble hydrated MCit salts are precipitated. It is also not satisfactory for the efficient separation of mixtures of certain rare earths. For the isolation of macro quantities of the rare earths, complexones like NTA and EDTA are preferred because their metal complexes are all soluble. One drawback with these chelating agents, when used in conjunction with cation exchangers in the hydrogen-ion form, is the precipitation of the slightly soluble free acids as hydrogen ion is displaced from the resin. For this reason, the columns are normally pre-conditioned with ions such as Fe^{3+} or Cu^{2+} which form stable, soluble complexes with the complexone. These function as retaining ions in the sense that, when the rare-earth ion has been displaced from the resin, they compete with it for the complexone and effectively promote its resorption by the resin.

Answers to problems

1.1. (*i*) The four nitrogens; (*ii*) the atoms of the imidodiacetic acid group; (*iii*) nitrogen and selenium; (*iv*) five oxygens of the carboxyl groups and three nitrogens; (*v*) two oxygens.

1.2. The NTA complex is more stable for each metal ion because it contains three chelate rings compared with two in the case of the IDA complex.

1.3. For a given number of oxygens on the ligand, the chain polyphosphate forms a more stable complex than the cyclic polymer. The complexes with cyclic polyphosphates are not very stable, primarily because of the steric impossibility of coordinating more than a few oxygens to one calcium ion.

2.1. Pb^{2+}, Hg^{2+}, and Cd^{2+}.

2.2. Trigonal bipyramidal, square planar, and trigonal bipyramidal.

2.3. Four-coordination with ethylenediamine is preferred in square-planar $\bar{C}u(en)_2^{2+}$. lg K_3 is small, probably because the third ligand coordinates in a unidentate fashion. Coordination as a bidentate would necessitate a major stereochemical rearrangement around copper. In the case of bipyridyl, steric factors prevent square-planar coordination of two molecules and less rearrangement is required for attachment of a third bidentate ligand.

2.4. 14.33.

2.5. The acidity should be 4 mol dm^{-3} or above.

3.1. Three geometrical isomers with *cis* X groups. These isomers have respectively two A groups, two B groups, and one A and one B group coplanar with the two Xs. All three have optical isomers. There are also two geometrical isomers with *trans* X groups.

3.2. Five containing (+)-pn and five containing (−)-pn.

3.3. Three geometrical isomers are possible. Two of these have *cis* B groups and are potentially resolvable.

4.1. For $Zn(NH_3)_2^{2+}$, $\Delta S^0 = +1.9$ J deg^{-1} mol^{-1}.
For $Zn(en)^{2+}$, $\Delta S^0 = +25.1$ J deg^{-1} mol^{-1}. The latter value shows the significant entropy increase which accompanies chelation.

4.2. $\Delta G_1^0 = -42.8$ kJ mol^{-1}; $\Delta G_2^0 = -36.2$ kJ mol^{-1}; $\Delta G_3^0 = 25.2$ kJ mol^{-1}.

$\Delta S_1^0 = +17.3$ J K^{-1} mol^{-1}; $\Delta S_2^0 = -7.3$ J K^{-1}; $\Delta S_3^0 = -51.6$ J K^{-1} mol^{-1}.

4.3. $\Delta H_1 = -41.27$ kJ mol^{-1}; $\Delta H_2 = -35.05$ kJ mol^{-1}.
At 293 K, $\Delta S_1 = -18.3$ J K^{-1} mol^{-1} and $\Delta S_2 = -34.6$ J K^{-1} mol^{-1}.
At 313 K $\Delta S_1 = -18.1$ J K^{-1} mol^{-1} and $\Delta S_2 = -33.67$ J K^{-1} mol^{-1}.

5.1. [Ca] = 10^{-7} mol dm^{-3}; [Mn] = 10^{-8} mol dm^{-3}.

5.2. 6×10^4 and 4.04×10^8.

5.3. [Ca] = 101.6 p.p.m.; [Mg] = 6.5 p.p.m.

5.4. [Cu] = 1.83 g l^{-1}; [Zn] = 1.05 g l^{-1}.

5.5. 0.995 V.

6.1. The stability decrease is parallel with an increase in ionic radius of the metal ion and indicates bonding is primarily electrostatic.

6.2. Three geometrical isomers are possible, designated respectively *cis-cis*, *cis-trans*, and *trans-cis*. The first designation refers to the relative orientation of the $-CF_3$ groups, the second to that of the acetyl groups of the trifluoroacetylacetonate ligands.

6.3. Cu(hfac)$_2$py$_2$. The electron-withdrawing power of the trifluoromethyl group is greater than that of the methyl group. As a result Cu(hfac)$_2$ is the stronger Lewis acid and forms the more stable adduct with pyridine.

6.4. $C_{15}H_{19}AlN_2O_{10}$. Two of the three chelate rings are substitued by $-NO_2$ in the 3-position.

6.5. $n = 3$.

7.1. Zinc is preferentially complexed by the complexones containing nitrogen and oxygen donors. Mercury is much more strongly bound by ligands which contain sulphur donors as well. The ineffectiveness of EDTA and DTPA in removing mercury suggests this metal is at least partly bonded by sulphur in haemoglobin.

7.2. A is $C_{14}H_{22}N_5O_4Co$. B is $C_9H_{19}N_4O_5Co$.
A is the pyridine adduct of the methylcobaloxime, CH$_3$CO-(C$_8$H$_{14}$N$_4$O$_4$). B is the aquo adduct of the same complex. In both, cobalt(I) is octahedrally coordinated.

Bibliography

General references

CHABAREK, S. and MARTELL, A.E. (1959). *Organic sequestering agents*. John Wiley and Sons, New York.
The chemical behaviour and applications of metal chelates in aqueous solutions.

DWYER, F. P. and MELLOR, D. P. (1964). *Chelating agents and metal chelates*. Academic Press, New York and London.
A reference work for senior students and research workers. Also of interest to biologists and medical scientists.

NAKAMOTO, K. and McCARTHY, P. J. (1968). *Spectroscopy and structure of metal chelate compounds*. John Wiley and Sons, New York.

PERRIN, D. D. (1970). *Masking and demasking of chemical reactions*. John Wiley and Sons, New York.

Chapter 1

HOLM, R. H., EVERETT, G. W., and CHAKRAVORTY, A. (1966). *Prog. inorg. Chem.* **7**, 83.
Metal complexes of Schiff bases and β-ketoamines.

McWHINNIE, W. R. and MILLER, J. D. (1969). *Adv. inorg. Radiochem.* **12**, 135.
The chemistry of complexes containing 2,2'-bipyridyl, 1,10-phenanthroline, or 2,2',6',2"-terpyridyl as ligands.

SCHWARZENBACH, G. (1961). *Adv. inorg. Radiochem.* **3**, 257.
General, selective, and specific formation of complexes by metallic cations.

Chapter 2

AHRLAND, S., CHATT, J., and DAVIES, N. R. (1958). *Q. Rev. chem. Soc.* **12**, 265.
The relative affinities of ligand atoms for acceptor molecules and ions.

LIPPARD, S. L. (1967). *Prog. inorg. Chem.* **8**, 109.
Eight-coordination chemistry.

MUETTERTIES, E. L. and SCHUNN, R. A. (1966). *Q. Rev. chem. Soc.* **20**, 245.
Pentacoordination.

Chapter 3

GILLARD, R. D. and IRVING, H. M. (1965). *Chem. Rev.* **65**, 603.
Conformational aspects of chelate rings.

KAUFFMAN, G. B. (1974). *Coord. Chem. Rev.* **12**, 105.
Alfred Werner's research on optically active coordination compounds.

Chapter 4

BEECH, G. (1969). *Q. Rev. chem. Soc.* **23**, 410.
Recent studies on the thermodynamics of metal complex formation.

JONES, G. R. H. and HARROP, R. (1973). *J. inorg. nucl. Chem.* **35**, 173.
Standard states and the chelate effect.

MARTELL, A. E. (1967). The chelate effect. In The Werner Centennial
Volume, *Adv. Chem. Ser.* **62**, 272-94.

Chapter 5

PRIBIL, R. (1972). *Analytical applications of EDTA and related compounds.* Pergamon Press, Oxford.

RINGBOM, A. (1963). *Complexation in analytical chemistry.* John Wiley
and Sons, New York.
A guide for the critical selection of analytical methods based on complexation reactions.

SCHWARZENBACH, G. (1957). *Complexometric titrations* (translated
by H. M. Irving). Methuen, London.

WEST, T. S. (1969). *Complexometry with EDTA and related reagents.*
B.D.H. Chemicals Ltd., Poole.

Chapter 6

von AMMON, R. and FISCHER, R. D. (1972). *Angew. Chem. Int. Ed.*
11, 675.
Applications of n.m.r. shift reagents.

FERNANDO, Q. (1965). *Adv. inorg. Radiochem.* **7**, 185.
Reactions of chelated ligands.

SINHA, A. P. B. (1971). *Spectrosc. inorg. Chem.* **2**, 255.
Review of fluorescence and laser action in rare-earth chelates.

THOMPSON, D. W. (1971). *Struct. bond.* **9**, 27.
Structure and bonding in inorganic derivatives of β-diketones.

Chapter 7

BROWN, D. G. (1973). *Prog. inorg. Chem.* **18**, 177.
Chemistry of vitamin B_{12} and related inorganic model systems.

CHOCK, P. B. and TITUS, E. O. (1973). *Prog. inorg. Chem.* **18**, 287.
Alkali-metal ion transport and biochemical activity.

HUGHES, M. N. (1972). *The inorganic chemistry of biological processes.* John Wiley and Sons, New York.
A survey of the occurrence and roles of metal ions of biological importance including a description of how the functions of these ions may be studied experimentally and also structural aspects of many biologically important chelates.

WILLIAMS, D. R. (1972). *Chem. Rev.* **72**, 203.
Metals, ligands, and cancer.

WILLIAMS, R. J. P. (1970). *Q. Rev. chem. Soc.* **24**, 331.
The biochemistry of sodium, potassium, magnesium, and calcium.

Chapter 8

SPEDDING, F. H. and DAANE, A. H. (1961). *The rare earths*. John Wiley and Sons, New York.

Index

Periodic Table

1A	IIA	IIIA	IVA	VA	VIA	VIIA	VIII			IB	IIB	IIIB	IVB	VB	VIB	VIIB	O
$_1$H 1·008																	$_2$He 4·003
$_3$Li 6·941	$_4$Be 9·012											$_5$B 10·81	$_6$C 12·01	$_7$N 14·01	$_8$O 16·00	$_9$F 19·00	$_{10}$Ne 20·18
$_{11}$Na 22·99	$_{12}$Mg 24·31											$_{13}$Al 26·98	$_{14}$Si 28·09	$_{15}$P 30·97	$_{16}$S 32·06	$_{17}$Cl 35·45	$_{18}$Ar 39·95
$_{19}$K 39·10	$_{20}$Ca 40·08	$_{21}$Sc 44·96	$_{22}$Ti 47·90	$_{23}$V 50·94	$_{24}$Cr 52·00	$_{25}$Mn 54·94	$_{26}$Fe 55·85	$_{27}$Co 58·93	$_{28}$Ni 58·71	$_{29}$Cu 63·55	$_{30}$Zn 65·37	$_{31}$Ga 69·72	$_{32}$Ge 72·59	$_{33}$As 74·92	$_{34}$Se 78·96	$_{35}$Br 79·90	$_{36}$Kr 83·80
$_{37}$Rb 85·47	$_{38}$Sr 87·62	$_{39}$Y 88·91	$_{40}$Zr 91·22	$_{41}$Nb 92·91	$_{42}$Mo 95·94	$_{43}$Tc 98·91	$_{44}$Ru 101·1	$_{45}$Rh 102·9	$_{46}$Pd 106·4	$_{47}$Ag 107·9	$_{48}$Cd 112·4	$_{49}$In 114·8	$_{50}$Sn 118·7	$_{51}$Sb 121·8	$_{52}$Te 127·6	$_{53}$I 126·9	$_{54}$Xe 131·3
$_{55}$Cs 132·9	$_{56}$Ba 137·3	$_{57}$La 138·9	$_{72}$Hf 178·5	$_{73}$Ta 180·9	$_{74}$W 183·9	$_{75}$Re 186·2	$_{76}$Os 190·2	$_{77}$Ir 192·2	$_{78}$Pt 195·1	$_{79}$Au 197·0	$_{80}$Hg 200·6	$_{81}$Tl 204·4	$_{82}$Pb 207·2	$_{83}$Bi 209·0	$_{84}$Po (210)	$_{85}$At (210)	$_{86}$Rn (222)
$_{87}$Fr (223)	$_{88}$Ra 226·0	$_{89}$Ac (227)															

Lanthanides	$_{57}$La 138·9	$_{58}$Ce 140·1	$_{59}$Pr 140·9	$_{60}$Nd 144·2	$_{61}$Pm (147)	$_{62}$Sm 150·4	$_{63}$Eu 152·0	$_{64}$Gd 157·3	$_{65}$Tb 158·9	$_{66}$Dy 162·5	$_{67}$Ho 164·9	$_{68}$Er 167·3	$_{69}$Tm 168·9	$_{70}$Yb 173·0	$_{71}$Lu 175·0
Actinides	$_{89}$Ac (227)	$_{90}$Th 232·0	$_{91}$Pa 231·0	$_{92}$U 238·0	$_{93}$Np 237·0	$_{94}$Pu (242)	$_{95}$Am (243)	$_{96}$Cm (248)	$_{97}$Bk (247)	$_{98}$Cf (251)	$_{99}$Es (254)	$_{100}$Fm (253)	$_{101}$Md (256)	$_{102}$No (254)	$_{103}$Lw (257)

SI units

Physical quantity	Old unit	Value in SI units
energy	calorie (thermochemical)	4·184 J (joule)
	*electronvolt—eV	$1·602 \times 10^{-19}$ J
	*electronvolt per molecule	96·48 kJ mol^{-1}
	erg	10^{-7} J
	*wave number—cm^{-1}	$1·986 \times 10^{-23}$ J
entropy (S)	eu = cal g^{-1} °C^{-1}	4184 J kg^{-1} K^{-1}
force	dyne	10^{-5} N (newton)
pressure (P)	atmosphere	$1·013 \times 10^{5}$ Pa (pascal), or N m^{-2}
	torr = mmHg	133·3 Pa
dipole moment (μ)	debye—D	$3·334 \times 10^{-30}$ C m
magnetic flux density (H)	*gauss—G	10^{-4} T (tesla)
frequency (ν)	cycle per second	1 Hz (hertz)
relative permittivity (ε)	dielectric constant	1
temperature (T)	*°C and °K	1 K (kelvin); 0 °C = 273·2 K

(* indicates permitted non-SI unit)

Multiples of the base units are illustrated by length

fraction	10^9	10^6	10^3	1	(10^{-2})	10^{-3}	10^{-6}	10^{-9}	(10^{-10})	10^{-12}
prefix	giga-	mega-	kilo-	metre	(centi-)	milli-	micro-	nano-	(*ångstrom)	pico-
unit	Gm	Mm	km	m	(cm)	mm	μm	nm	(*Å)	pm

The fundamental constants

Avogadro constant	L or N_A	$6·022 \times 10^{23}$ mol^{-1}
Bohr magneton	μ_B	$9·274 \times 10^{-24}$ J T^{-1}
Bohr radius	a_0	$5·292 \times 10^{-11}$ m
Boltzmann constant	k	$1·381 \times 10^{-23}$ J K^{-1}
charge of a proton	e	$1·602 \times 10^{-19}$ C
(charge of an electron $= -e$)		
Faraday constant	F	$9·649 \times 10^{4}$ C mol^{-1}
gas constant	R	$8·314$ J K^{-1} mol^{-1}
nuclear magneton	μ_N	$5·051 \times 10^{-27}$ J T^{-1}
permeability of a vacuum	μ_0	$4\pi \times 10^{-7}$ H m^{-1} or N A^{-2}
permittivity of a vacuum	ε_0	$8·854 \times 10^{-12}$ F m^{-1}
Planck constant	h	$6·626 \times 10^{-34}$ J s
(Planck constant)/2π	$\hbar$	$1·055 \times 10^{-34}$ J s
rest mass of electron	m_e	$9·110 \times 10^{-31}$ kg
rest mass of proton	m_p	$1·673 \times 10^{-27}$ kg
speed of light in a vacuum	c	$2·998 \times 10^{8}$ m s^{-1}

$\ln 10 = 2·303$ $\quad$ $\ln x = 2·303 \lg x$ $\quad$ $\lg e = 0·4343$ $\quad$ $\pi = 3·142$
$R \ln 10 = 19·14$ J K^{-1} mol^{-1} $\quad$ $RTF^{-1} \ln 10 = 59·16$ mV at 298·2 K